Wondrous One Sheet Origami

Wondrous One Sheet Origami

Meenakshi Mukerji

CRC Press
Taylor & Francis Group
Boca Raton   London   New York

CRC Press is an imprint of the
Taylor & Francis Group, an **informa** business

AN A K PETERS BOOK

Cover photos:
    Front - Fractal Sakura
    Back - Elm Leaf (top) and Abstract Flowers (bottom)

Overleaf: Kusumita Butterfly

CRC Press
Taylor & Francis Group
6000 Broken Sound Parkway NW, Suite 300
Boca Raton, FL 33487-2742

© 2020 by Taylor & Francis Group, LLC
CRC Press is an imprint of Taylor & Francis Group, an Informa business

No claim to original U.S. Government works

Printed on acid-free paper

International Standard Book Number-13: 978-0-367-20813-4 (Hardback)
                              978-0-367-20810-3 (Paperback)

Visit the Taylor & Francis Web site at
http://www.taylorandfrancis.com

and the CRC Press Web site at
http://www.crcpress.com

To All Who Inspire Me

Contents

Foreword

Meenakshi Mukerji is one of today's masters of modular origami, geometric designs comprised of multiple folded pieces of paper assembled without glue. In her newest book, she brings her ingenuity and creativity to designs made from a single piece of paper. Among the most appealing aspects of this book is the way she subtly manipulates a purely geometric design to form a flower (including a beautiful 12-petaled Hollyhock), a leaf, a butterfly, or familiar symbols such as the four suits from a deck of cards. Although many of her designs are complex and three-dimensional, requiring an abundance of both patience and dexterity to fold, she provides crystal-clear diagrams with lucid text and doesn't stint on providing perspective drawings (which are challenging and time-consuming to produce) when needed. In addition to presenting her own stunning designs, Mukerji offers up designs and biographies of three "guest folders" whose work is as stunning as her own. Of these designs, my favorite is the mind-boggling Floral Perpetua, which uses a recursive (self-repeating) folding process to mimic the concentric layers of petals in a blooming flower. It's worth the price of admission all on its own. Like this spectacular model, Mukerji's book yields riches in perpetuity!

Peter Engel, author of *Origami Odyssey: A Journey to the Edge of Paperfolding* [Eng11], *10-Fold Origami* [Eng16], and *Folding the Universe: Origami from Angelfish to Zen* [Eng94]

Preface

After nearly 15 years of designing modular origami and writing four books on the subject, I am excited to present my very first book on single sheet designs. I have folded and enjoyed single sheet designs in the past, but it is only lately that I have delved into it deeper. Such amazing models as Shuzo Fujimoto's Hydrangea [Fuj10] and Roman Diaz's [Dia06] Fractal Flower [Dia12] renewed my interest in this original form of the ancient art. More recently, John Montroll's book *Galaxy of Origami Stars* [Mon12] completely enraptured me with the single sheet genre. The challenges of one sheet designs are intriguing and refreshing after having designed modular origami for so long. Since my name is somewhat associated with modular designs, I worked the words "one sheet" into the title so no presumptions would be made by folders.

Not only are the designs presented here made from a single sheet, they also follow the basic rules of origami, i.e., they are from single squares with no cuts or glue. There are a few designs that begin with other polygons, but they still follow the same rules once the polygon, the starting medium, has been obtained by cutting. As different as these new designs might be from my previous ones, my love for symmetry and geometry still remain noticeably unscathed. Notably, most of the designs in this book incorporate color change (both sides of the paper visible in the final model), so duo paper, i.e., paper with two different colors on each side, is recommended for the best outcome.

The book starts with some origami basics that are customarily included in any origami book. If you are already familiar with these basics, please feel free to skip ahead. The general structure of the book follows a natural progression starting with simple designs and gradually going into more complex ones, although there may be a few minor exceptions based on context. I always like to include designs of new and upcoming artists, so I have presented creations of guest folders as well. It is my privilege to introduce them to origami enthusiasts around the world, if one is not already familiar with them. A large number of artists that I have included in my books have gone on to write their own, giving me a huge sense of fulfillment.

The diagrams in this book follow standard origami symbols pioneered by origami master Akira Yoshizawa [Yos84]. The styles of guest contributors have been preserved instead of changing them simply for the sake of consistency across the book. The origami diagramming language is very powerful with barely any need for written instructions with some exceptions when a little explanation may be required. To keep the diagrams simple and uncluttered, I have refrained from showing layers except when necessary. Some crease lines have been omitted at times for the sake of clarity, without compromising the outcomes.

Wondrous One Sheet Origami covers a wide range of folding levels from simple to high intermediate, with more emphasis on the latter. The book should appeal to audiences 12 years of age and above or to children folding at higher than age level. Most of the designs are flat and hence suitable for mounting on greetings cards or for framing as gifts. My understanding from social media is that the designs here are quite sought after. So, I am confident that you will enjoy the book because I am proceeding with useful feedback from origami enthusiasts across the globe. Sit back, relax, and enjoy while you explore the huge potential of a single sheet of paper!

Cupertino, California
December 2013

Second Edition Addendum

It is my pleasure to release a new edition of this book, six years later, and to publish with CRC Press. I am excited to present 40% new material, including some leaves that are my absolute favorites. You will also find new simple designs, recursive designs, flowers, and boxes. With compact formatting, you will get about 200 pages worth of goodies squeezed into just a 150-page book without loss of clarity and artfulness, and while keeping it economical. Once again, I hope you enjoy folding from this new edition of the book which has been a labor of love.

Cupertino, California
May 2019

Acknowledgments

While I have designed, folded and diagrammed most of the origami in this book, and done much of the book layout and photography, it would not have been complete without the invaluable help of the people named here. I would like to express my deepest gratitude to them all.

First of all, I would like to thank my guest contributors, Evan Zodl (NJ), Dáša Ševerová (Slovakia), Christiane Bettens (Switzerland), and David Donahue (CA), for being generous with their beautiful designs and allowing me to include them in this book. They have been extremely cooperative with the challenges of importing and exporting diagrams between formats. They already have a huge fan base, but I hope that with this book I am able to introduce them to an even wider audience.

Thanks to Roman Diaz (Uruguay), Endre Somos (Hungary), Jorge Jaramillo (Colombia), and Hitoshi Fujimoto (son of Shuzo Fujimoto, Japan). They were immensely kind to let me include in this book designs that are based on theirs. Further details are presented in their respective chapters.

I would like to thank Kedar Amladi (CA) for patiently taking photographs using his innovative homemade diffused light apparatus, and for processing the photos as well. He was then a sophomore at Carnegie Mellon University, and I was lucky that he was available during a break. Thanks also to Rosalinda Sanchez (AZ) for enhancing the cover photograph. Thanks to Sara Adams (Germany) [Ada07] for patiently making video instructions for some of the designs in this book.

I am fortunate to have a brilliant team of testers who patiently tested and pointed out errors easy to overlook. They came up with wonderful suggestions for improvement as well. I would like to thank my tester-folders: Rui Roda (Portugal), JC Nolan (CA), Rebecca Harris (CT), Scott Cramer (NH), and Charul Patil (India). Rui has been the power tester, test-folding a large part of the book. JC came up with meaningful suggestions to improve my Fractal Sakura diagrams, and helped develop a crease pattern for it. Scott tested many of the leaves for me.

Special thanks to Mark Kennedy (PA), Hilli Zenz (Germany), and JC Nolan for the beautiful handmade paper that they generously gave me. I have used the paper to fold some of the designs for very nice effects. As usual, Vinita Singhal (CA) has been my taken-for-granted technical adviser, and Jean Jaiswal (OR) my resourceful proofreader. My heartfelt thanks go to both of them. Thanks to my copy editor, Charlotte Byrnes (MA), for her patience and her keen eye for the minutest of errors that one may easily overlook. Thanks to Callum Fraser and the rest of the team at CRC Press, and Arun Kumar of Nova Techset for making the book possible. And last but not least, I would like to thank everyone who inspired me including my family, friends, visitors of my website, and fans of my origami.

An arrangement of Abstract Flowers (p. 88) folded with hand colored harmony paper. Leaf diagrams not included.

Photo Credits

I would like to credit and thank the following profoundly for folding, photography, or both:

◈ Teresa Montero-Cañibe: Flower arrangement, folding and photo on p. xiv.

◈ Billie Griebler: Wreath arrangement, folding and photo on p. xvi.

◈ Kedar Amladi: Photos of Butterfly on title page; Mum, Hollyhock, and Marigold on p. 29; Hollyhock on p. 35; Kusumita Flowers and Leaves on p. 40; Butterflies on p. 41; and Fractal Sakura Finish 2 on p. 108.

◈ Rui Roda: Folding and photos of Simple Swan on p. 22 (bottom), Butterflies on p. 56, 13-petalled Gaillardia on p. 95 (bottom right).

◈ JC Nolan: Folding of Fractal Sakura Base on p. 108 (top row).

◈ Evan Zodl: Folding and photos of 12-Pointed EZ Star and Star Tower on p. 120 (first two rows).

◈ Dáša Ševerová: Folding and photos of Floral Perpetua on p. 120 and Twisted Boxes on p. 131 (top).

◈ Christian Bettens: Folding and photos of Leafy Boxes on p. 131 (bottom) and Four-Leaf Tato Boxes on p. 137.

◈ Sara Adams: Photo of Hydrangea Box on p. 139.

◈ Indra Singhal: Photo of author on p. 140.

◈ Hilli Zenz: Photo of Dáša Ševerová on p. 142.

◈ All other folding and photos are by the author.

Wreath made with Poinsettia (p. 39), Elm Leaves (p. 71). Other flowers, not diagrammed in book, may be found in author's prior book, *Origami All Kinds* [Muk17].

Traditional Lily, Five Petal Lily, Tuberose and Oleander, please see Chapter 2.
(Traditional leaf diagram not included).

1 ◈ Origami Basics

The word *origami* is based on two Japanese words: *oru* (to fold) and *kami* (paper). Although this ancient art of paper folding started in Japan and China, origami is now a household term around the world. Most people have probably folded at least a paper boat or an airplane at some point in their lives. However, origami has evolved immensely over time and is much more than just a handful of traditional models. That it is child's play is quite a misconception; there are more and more people from a wide range of backgrounds immersing themselves into the joys of origami because it fulfills a certain craving for a unique kind of creativity that brings together both sides of the brain.

The origami symbols used here are standard. The symbols were pioneered by renowned origami master Akira Yoshizawa (1911–2005) [Yos84] and further developed by more recent creators, adding more symbols and terms leading to the current standard. Origami diagramming follows the convention that the execution of a current step results in what is diagrammed in the next step. Different artists may have slightly different ways of representing the same symbols essentially to mean the same moves. Usually every origami book will have their own glossary of symbols just in case there are any doubts or someone is folding for the first time.

I would like to emphasize precision in folding. It is the number one secret to successful finished models pleasing to the eyes. So, I would recommend that you fold as precisely as possible, slowing down instead of rushing. After all, origami is for relaxation, so why rush? Make sure that your work area is well lit so you can see the reference points and lines clearly. Oblique lighting is often more helpful than overhead lighting. If you cannot understand a step, looking ahead to the next step helps immensely.

Origami Paper and Tools

Origami can be folded from practically any type of paper. But most designs have some paper that works best for them, and mainly experience can tell you what to use. Some models might require sturdy paper, while some others might require paper that creases softly. Also, it is advisable to keep a small set of basic tools handy if you plan to do origami regularly.

Origami Tools

Creasing Tools. The most basic tool that is used in origami is a bone folder. It allows for making precise and crisp creases and prevents your nails or fingertips from becoming sore when folding excessively or folding many layers together. Substitutes are plastic cards or similar things.

Cutting Tools. Although cutting is prohibited in pure origami, cutting tools are required for the initial sizing of the paper. A great cutting tool would be a paper guillotine, but it is bulky and may not be readily accessible. I find a portable photo trimmer with replaceable blades to be a great substitute. They are inexpensive and easily carried anywhere. Scissors and blades may be used, but it is difficult to get straight cuts, especially if one is not experienced.

Other Tools. Tweezers, chopsticks, or stylus-like objects may be used to access hard-to-reach places such as in between layers of paper or to deal with paper that becomes too small during the folding process to maneuver with fingers. Chopsticks, screw drivers, or knitting needles are great for curling in designs that call for it.

Paper that you use need not necessarily be paper manufactured specifically for origami purposes. A vast range of papers from gift wrap to scrapbooking paper work very well. You just have to apply your imagination. Often times I reuse flier papers printed only on one side, which work well for designs that do not show both sides of the paper when finished. It's a great way to recycle and reuse or simply to practice folds.

Commonly Available Origami Paper

Kami. This is the most readily available paper sold for origami. It is solidly colored on one side and white on the other with a weight thinner than common printer paper.

Duo. Paper that has different colors on the front and back sides is advisable for designs with color change, i.e., when both sides of the paper are visible in the finished model.

Printer paper. Often called copy paper, this is paper that is white or colored (usually the same on both sides) and is commonly used in home or office computer printers.

Mono. Such paper has the same color on both sides. Printer paper is an example of mono paper, and it may be available in other weights including *kami* weight.

Harmony. This paper has some harmonious pattern formed by various colors or shades blending into one another. This paper can have magical effects on certain designs.

Chiyogami. Origami paper with patterns, usually small, printed on it.

Washi. This handmade Japanese paper has plant fiber in the pulp that gives it texture. The fibers are often visible, giving a beautiful effect.

Foil backed. These sheets have metallic foil on one side and paper on the other side.

Tant. This mono paper is of about printer paper weight and contains some texture.

Some other papers of choice of origami artists are Elephant Hide, Stardream, Unryu, Glassine, Lokta, and Kraft. It is beyond the scope of this book to discuss all of these, but the reader is encouraged to look them up on the internet and try them out. Many times it's the choice of paper that brings out the best in a design.

Many artists like to create their own paper. To obtain duo paper of desired colors not readily available, one may sandwich together two different colors of paper with suitable glue, such as methyl cellulose, or use paint to get the two desired colors. Tissue papers sandwiched with foil in between, known as tissue foil, is a favorite of many artists because it enables some amount of sculpting. Spray painting may be done for decoration or for a feel of texture. Some designs look stunning with textured paper. There is no limit to what can be done with the starting piece of paper to achieve dramatic effects on the finished model.

Paper recommendations have been made in the designs presented whenever relevant. When no recommendation is made, the choice is left to the reader. A few examples of some of the above-described paper used for folding in this book are tissue sandwich (p. iii); duo (flowers in Chapter 6); corona harmony (p. 28); printer paper (p. 29, top); Kraft (Shamrock 2, p. 40); gift wrap (printed butterflies, p. 41); tissue foil (p. 53); Dye Watercolor (Philodendron, p. 55); spray painted duo paper (p. 95, bottom right), Tant (p. 120, middle), and duo foil (p. 137, bottom right).

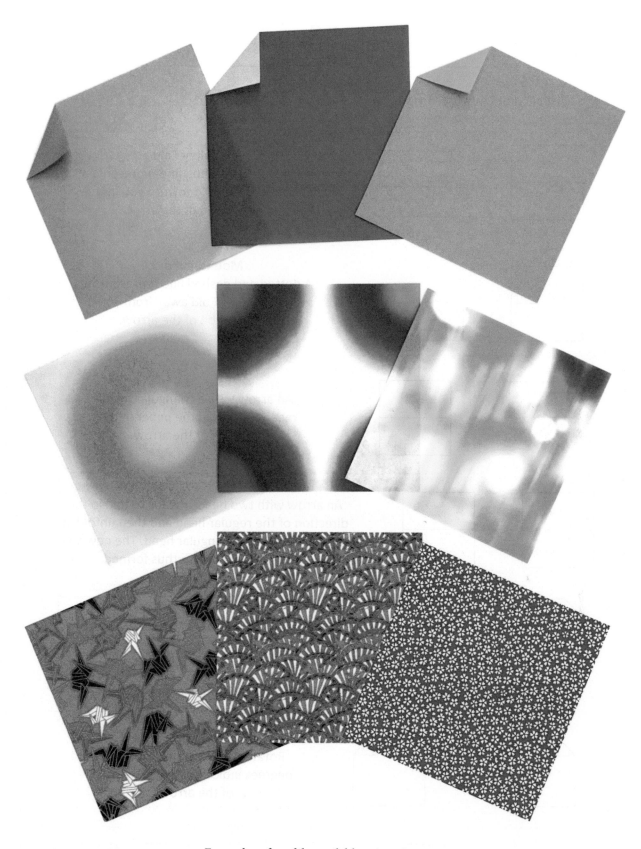

Examples of readily available origami paper:

Top: Mono, *Kami*, and Duo. Middle: Corona Harmony, Corona Harmony, and Dye Watercolor Origami Paper.
Bottom: Crane patterned Washi, Fan patterned Washi, and Chiyogami.

Origami Symbols

Below is a list of commonly used origami symbols.

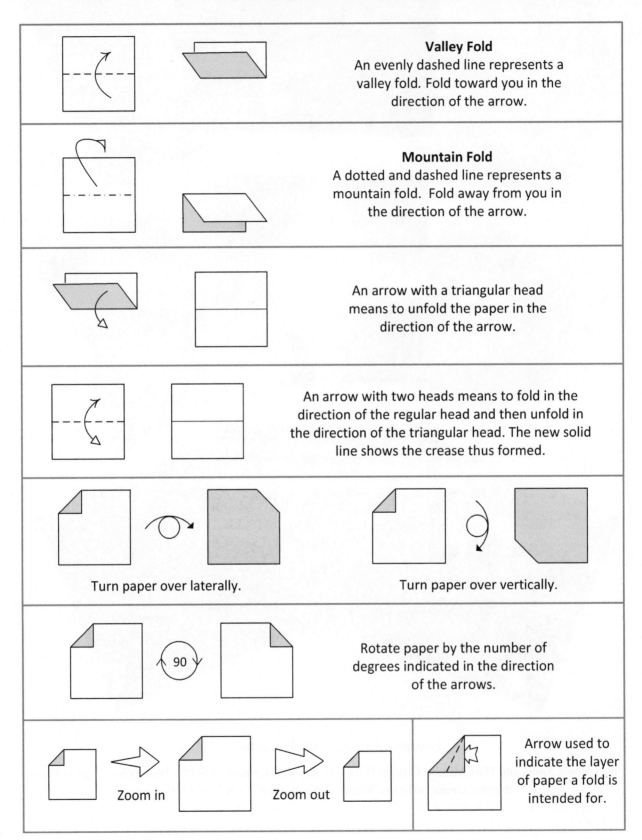

Valley Fold
An evenly dashed line represents a valley fold. Fold toward you in the direction of the arrow.

Mountain Fold
A dotted and dashed line represents a mountain fold. Fold away from you in the direction of the arrow.

An arrow with a triangular head means to unfold the paper in the direction of the arrow.

An arrow with two heads means to fold in the direction of the regular head and then unfold in the direction of the triangular head. The new solid line shows the crease thus formed.

Turn paper over laterally.

Turn paper over vertically.

Rotate paper by the number of degrees indicated in the direction of the arrows.

Zoom in

Zoom out

Arrow used to indicate the layer of paper a fold is intended for.

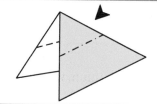 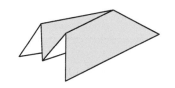

**Sink or Reverse Fold or
Inside Reverse Fold**
Push in the direction of the
arrow to arrive at the result.

Equal lengths.

Equal angles.

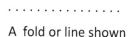

A fold or line shown
in X-ray vision, i.e., it
is behind layers.

Pull out paper.

Figure is truncated for
diagramming convenience.

Repeat once, twice, or as many
times as indicated by the tail of
the arrow.

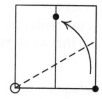

Fold to match dots using the circled point as pivot.

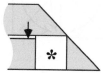 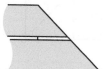

✳ : Tuck in opening underneath.

 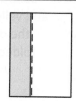

Fold repeatedly to
arrive at the result.

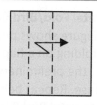 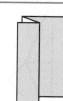   

Pleat Fold

An alternate mountain and valley fold to form a pleat. Two examples are shown.

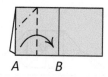 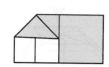

Squash Fold
Valley fold paper to the right
while making the mountain fold,
such that A meets B.

 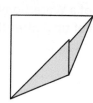

Rabbit Ear Fold
This applies to a triangular area of the
paper. Two of the angles are bisected
while aligning one side with another. The
result is a flap that looks like a rabbit ear.

Common Bases and Terms

Listed below are some common bases and terms in origami. Other bases, not so common, will be introduced as needed.

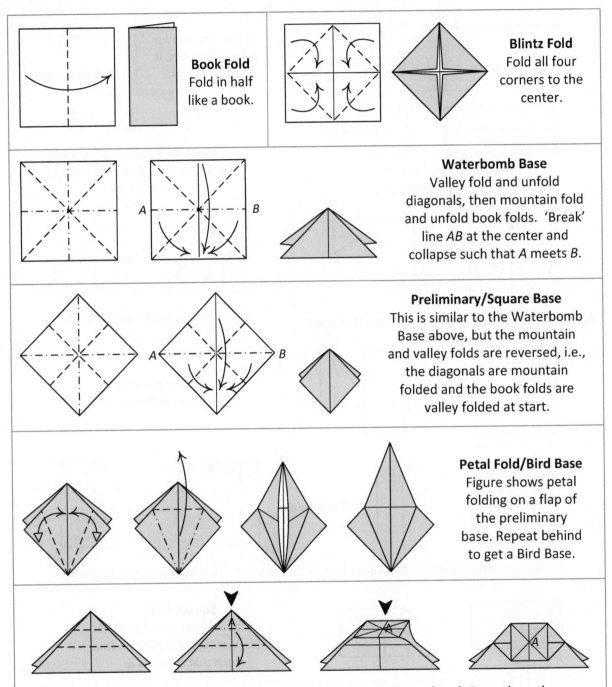

Book Fold
Fold in half like a book.

Blintz Fold
Fold all four corners to the center.

Waterbomb Base
Valley fold and unfold diagonals, then mountain fold and unfold book folds. 'Break' line *AB* at the center and collapse such that *A* meets *B*.

Preliminary/Square Base
This is similar to the Waterbomb Base above, but the mountain and valley folds are reversed, i.e., the diagonals are mountain folded and the book folds are valley folded at start.

Petal Fold/Bird Base
Figure shows petal folding on a flap of the preliminary base. Repeat behind to get a Bird Base.

Spread Squash: Figure shows spread squashing the tip of a Waterbomb Base along the two valley creases shown. Note the final position of tip *A* after the squash.

Folding a Square into Thirds

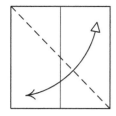 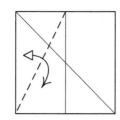 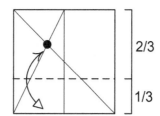 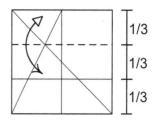

Crease book fold and one diagonal. Crease diagonal of left rectangle to find 1/3 point. Bring bottom edge to this point and top edge to new line.

Windmill Base

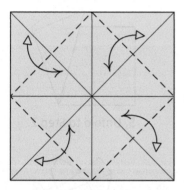 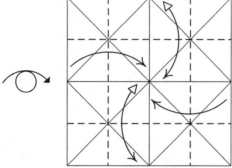 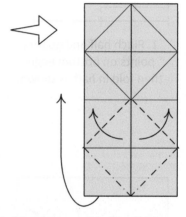

1. Crease diagonals and book folds. Then blintz fold and unfold.

2. Fold quarters horizontally and unfold. Then fold quarters vertically.

3. Mountain and valley fold on existing creases while bringing bottom edge to center.

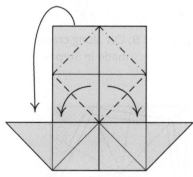 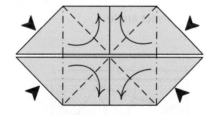 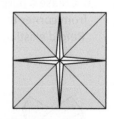

4. Repeat Step 3 on the top, bringing top edge to center.

5. Squash the four flaps.

Windmill Base

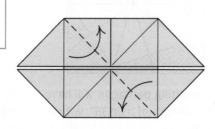 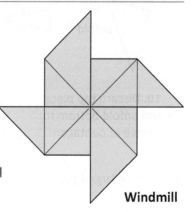

Note: In Step 5 if we just turn two of the flaps as shown, we get a Windmill and hence the name of the base.

Windmill

Pentagon from a Square

The method shown here is by Toyoaki Kawai [Kaw70]. There are various other methods available for obtaining a pentagon from a square, but I have chosen this one because of its ease and elegance of folding and paper efficiency.

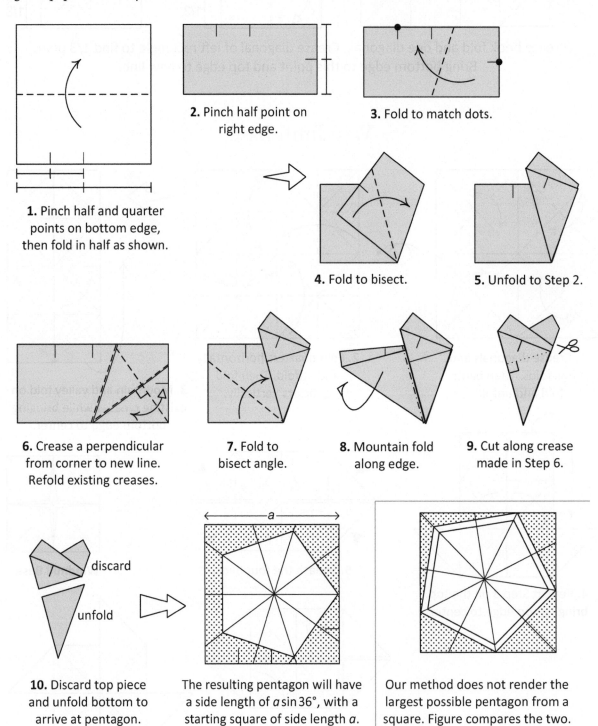

1. Pinch half and quarter points on bottom edge, then fold in half as shown.

2. Pinch half point on right edge.

3. Fold to match dots.

4. Fold to bisect.

5. Unfold to Step 2.

6. Crease a perpendicular from corner to new line. Refold existing creases.

7. Fold to bisect angle.

8. Mountain fold along edge.

9. Cut along crease made in Step 6.

10. Discard top piece and unfold bottom to arrive at pentagon.

The resulting pentagon will have a side length of $a \sin 36°$, with a starting square of side length a.

Our method does not render the largest possible pentagon from a square. Figure compares the two.

Folding the largest possible pentagon from a square is not straightforward and beyond the scope of this book. Please see [Mat01] if you are interested.

Hexagon from a Square

This method of folding a hexagon uses the traditional way of folding 30°/60° folds.

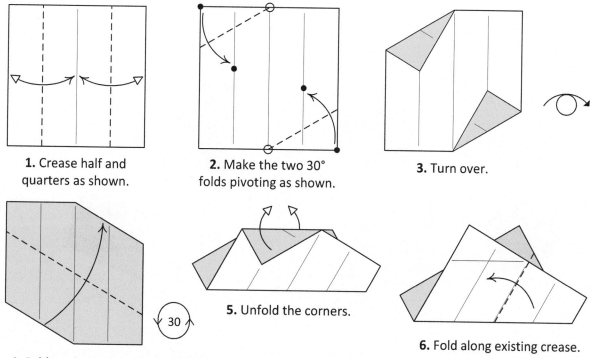

1. Crease half and quarters as shown.

2. Make the two 30° folds pivoting as shown.

3. Turn over.

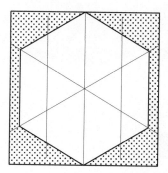

4. Fold to align edges. Note that the ends of edges will be offset. Rotate.

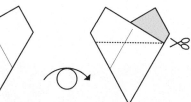

5. Unfold the corners.

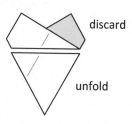

6. Fold along existing crease.

7. Fold along the edge.

8. Turn over.

9. Cut along existing crease.

10. Discard top. Unfold bottom.

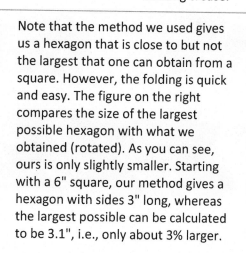

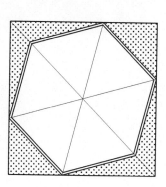

The resulting hexagon. Shaded portion is the discarded part.

Note that the method we used gives us a hexagon that is close to but not the largest that one can obtain from a square. However, the folding is quick and easy. The figure on the right compares the size of the largest possible hexagon with what we obtained (rotated). As you can see, ours is only slightly smaller. Starting with a 6" square, our method gives a hexagon with sides 3" long, whereas the largest possible can be calculated to be 3.1", i.e., only about 3% larger.

For understanding the largest possible hexagon from a square, please refer to [Gupta].

2 ◈ Simple Designs

Simple Hearts (p. 12), Simple Diamond (p. 11), and Ace of Diamonds (p. 13).

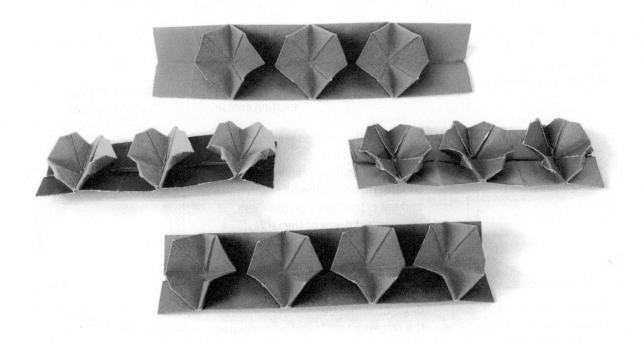

Flower Strips (p. 25), showing both variations, and with and without extra spaces on the side.

Simple Diamond

(Created 2013)

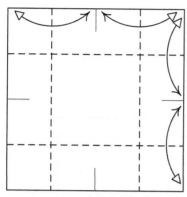

1. Pinch ends of book folds, then fold the quarters shown and unfold.

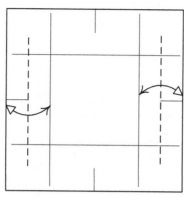

2. Fold and unfold as shown, making sure not to crease near the edges.

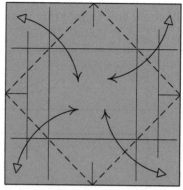

3. Blintz fold and unfold.

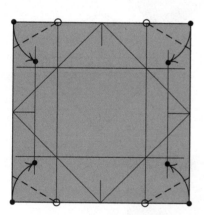

4. Fold corners to new creases.

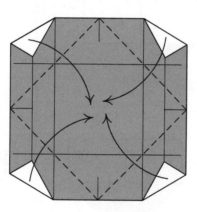

5. Re-crease blintz folds.

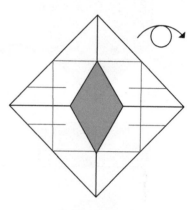

6. Turn over.

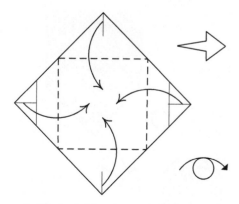

7. Blintz fold using existing creases, through all layers.

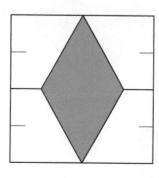

Simple Diamond

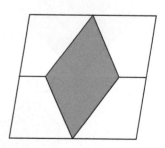

You can make Simple Diamond stand by adjusting the bottom flap at the back.

(Photo on p. 10.)

Simple Hearts

(Created 2009)

These hearts are good with leftover strips of paper, especially foil backed paper. The diagrams below illustrate how to fold three types of hearts with slightly different shapes.

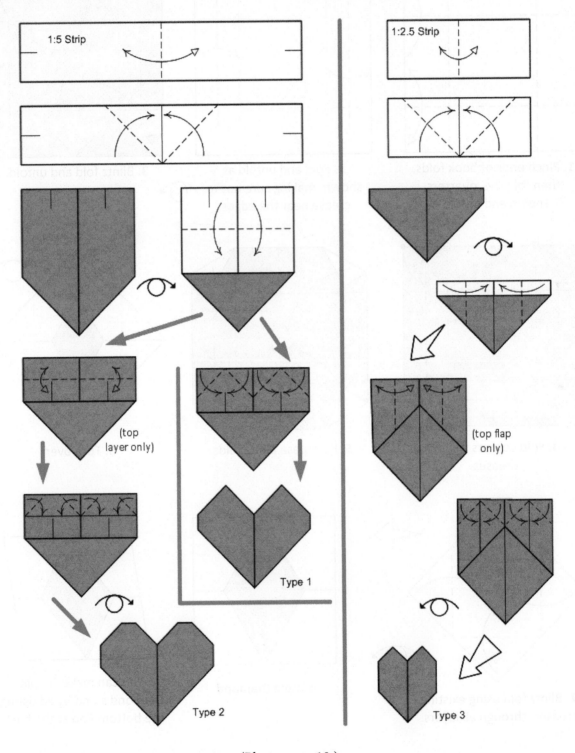

1:5 Strip

1:2.5 Strip

(top layer only)

(top flap only)

Type 1

Type 2

Type 3

(Photo on p. 10.)

Ace of Diamonds

(Created 2013)

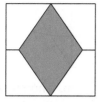

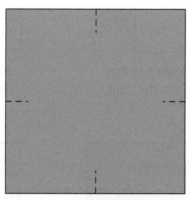

1. Pinch ends of both book folds.

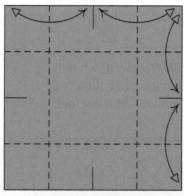

2. Cupboard fold and unfold both ways.

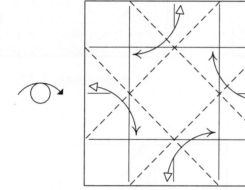

3. Fold and unfold the four diagonals shown.

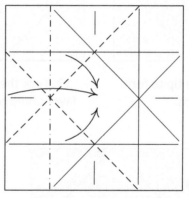

4. Collapse the left side using the existing valley and mountain creases.

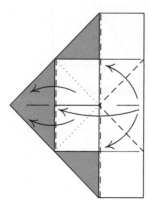

5. Valley fold white flaps to the left while repeating Step 4 on the right.

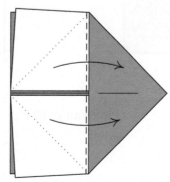

6. Valley fold flaps to the right while squashing the second layer getting pulled from behind.

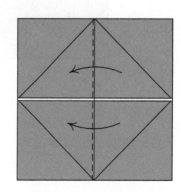

7. Fold flaps to the left.

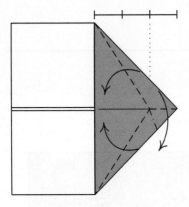

8. Narrow down red triangle by about a third with a rabbit ear fold.

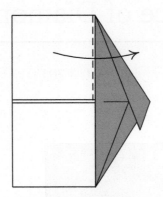

9. Bring top flap back to the right.

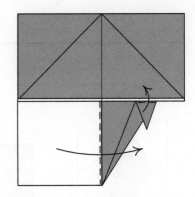

10. Fold tip up. Then turn bottom flap to the right.

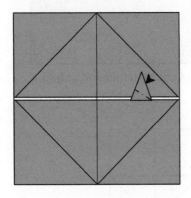

11. Squash the tip to form a latch.

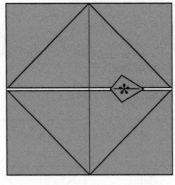

12. Tuck the latch in the layer underneath to hide.

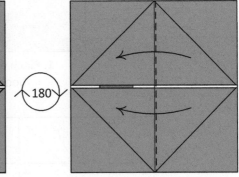

13. Repeat Steps 7–12.

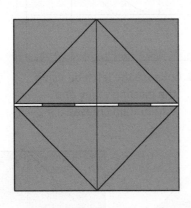

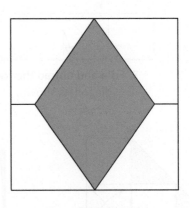

Ace of Diamonds

(Photo on p. 10.)

Traditional and Five Petal Lilies

The purpose of this exercise is to demonstrate how to map the folds of a square piece of paper on to a pentagonal one. With this knowledge you may be able to transpose many designs starting from a square into a pentagon or other polygon, resulting in different-looking finishes from the original model. Shown below is how to make a Traditional Lily from a square sheet of paper, which most of you probably already know, alongside with how to make a Five Petal Lily from a pentagonal sheet of paper.

Start with a square for the Traditional Lily and a pentagon for the Five Petal Lily. Please see p. 8 for instructions on how to make a pentagon.

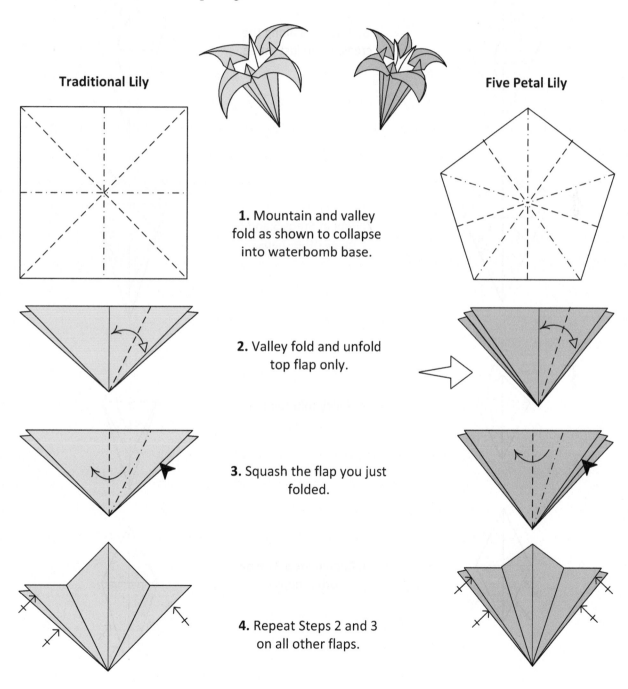

Traditional Lily

Five Petal Lily

1. Mountain and valley fold as shown to collapse into waterbomb base.

2. Valley fold and unfold top flap only.

3. Squash the flap you just folded.

4. Repeat Steps 2 and 3 on all other flaps.

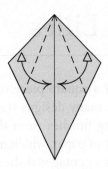

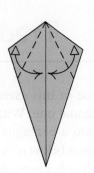

5. Valley fold and unfold top flap only.

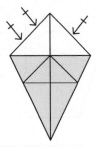

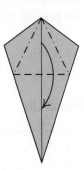

6. Petal fold along the creases you just made.

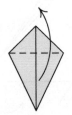

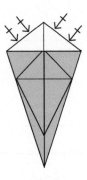

7. Repeat Steps 5 and 6 on all other flaps.

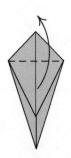

8. Valley fold top flap.

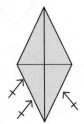

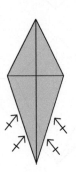

9. Repeat Step 8 on all other flaps.

Traditional and Five Petal Lilies

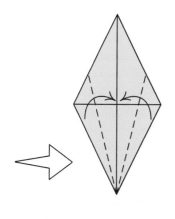

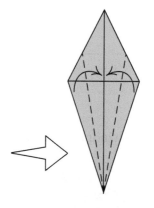

10. Valley fold edges of top flap to center.

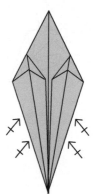

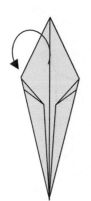

11. Repeat Step 10 on all other flaps.

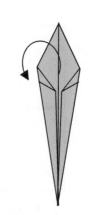

12. Curl front petal toward you. Repeat on all other petals to arrive at finished Lily.

Traditional Lily

Five Petal Lily

(Photo on p. xvi, left.)

Tuberose

(Created 2006)

Start with a windmill base (p. 7).

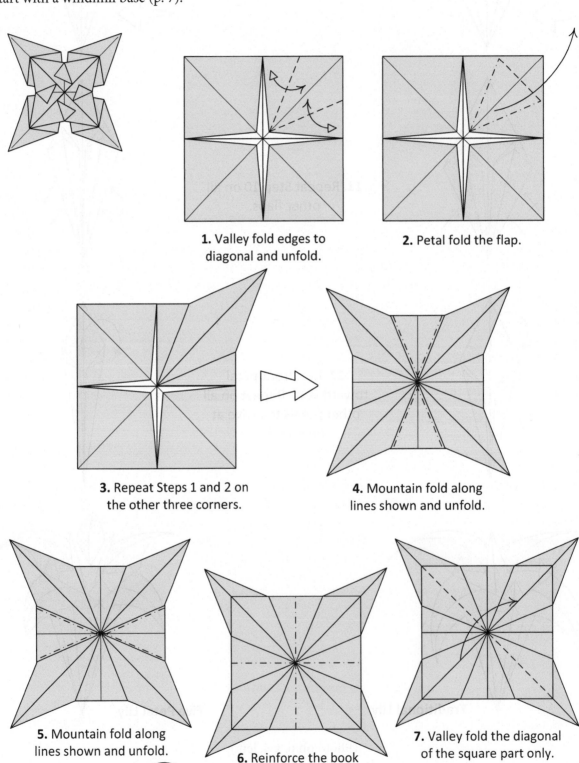

1. Valley fold edges to diagonal and unfold.

2. Petal fold the flap.

3. Repeat Steps 1 and 2 on the other three corners.

4. Mountain fold along lines shown and unfold.

5. Mountain fold along lines shown and unfold.

6. Reinforce the book folds as mountain folds through all layers.

7. Valley fold the diagonal of the square part only.

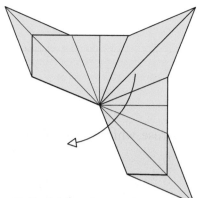

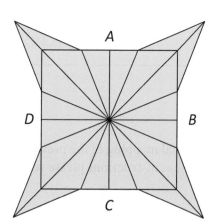

8. Unfold and repeat on the other diagonal.

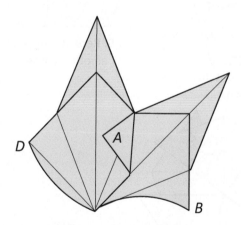

9. Pleat fold at *A* as shown.

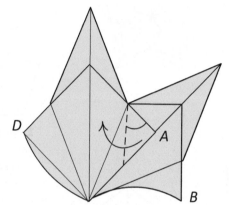

10. Fold tip *A* as shown such that the marked angle is approximately 45° or a tad bit larger. Crease firmly.

11. Repeat Steps 9 and 10 at *B*, *C*, and *D*. Earlier folds tend to unfold; repeat those folds if necessary until all four corners stay folded.

Curl the petals with a chopstick-like object for a more natural look.

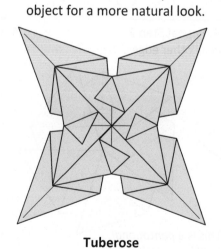

Tuberose

(Photo on p. xvi, right.)

Oleander

(Created 2006)

The Oleander design is a pentagonal mapping of the previous design, Tuberose, which is from a square. Start with a pentagon. For pentagon instructions please see p. 8.

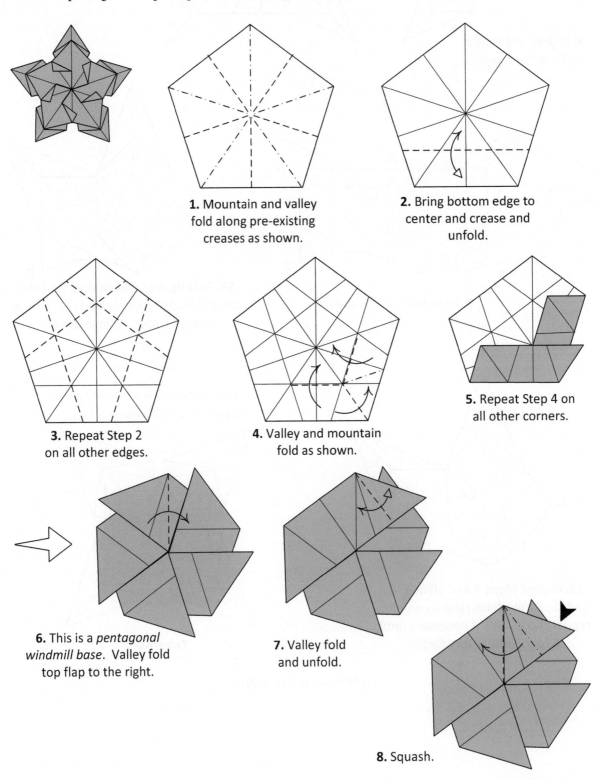

1. Mountain and valley fold along pre-existing creases as shown.

2. Bring bottom edge to center and crease and unfold.

3. Repeat Step 2 on all other edges.

4. Valley and mountain fold as shown.

5. Repeat Step 4 on all other corners.

6. This is a *pentagonal windmill base*. Valley fold top flap to the right.

7. Valley fold and unfold.

8. Squash.

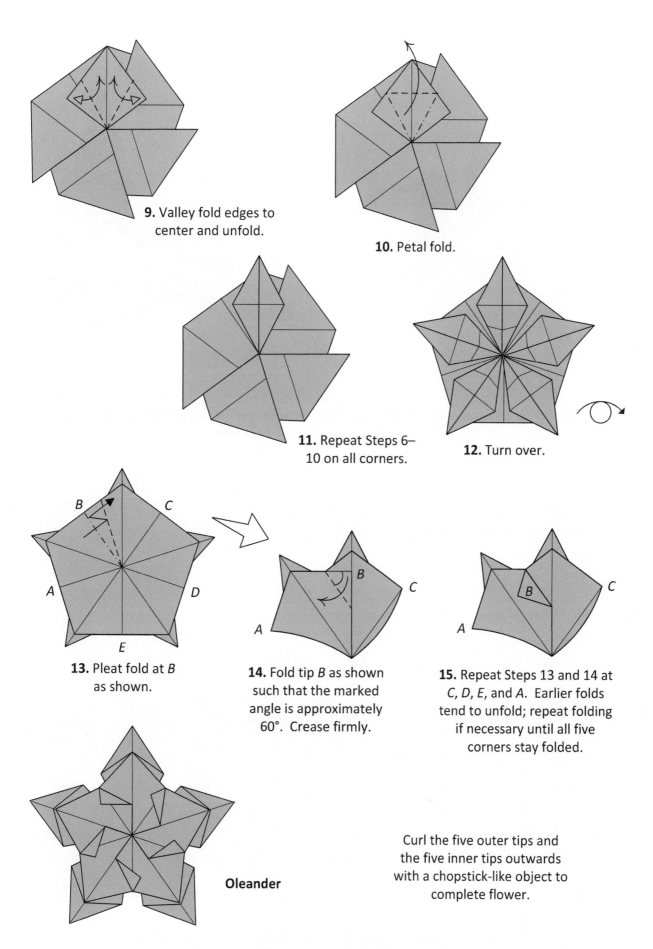

9. Valley fold edges to center and unfold.

10. Petal fold.

11. Repeat Steps 6–10 on all corners.

12. Turn over.

13. Pleat fold at *B* as shown.

14. Fold tip *B* as shown such that the marked angle is approximately 60°. Crease firmly.

15. Repeat Steps 13 and 14 at *C, D, E,* and *A.* Earlier folds tend to unfold; repeat folding if necessary until all five corners stay folded.

Oleander

Curl the five outer tips and the five inner tips outwards with a chopstick-like object to complete flower.

(Photo on p. xvi, right.)

Simple Peacock (p. 23) folded with harmony paper, and Simple Swan (p. 24) folded with black paper spotted with whiteout ink, giving the feel of a loon.

Swan with neck narrowed and a more intricate finish of the head and beak.

This idea by Rui Roda is being left as an exercise for the keener folder.

Simple Peacock

(Created 2017)

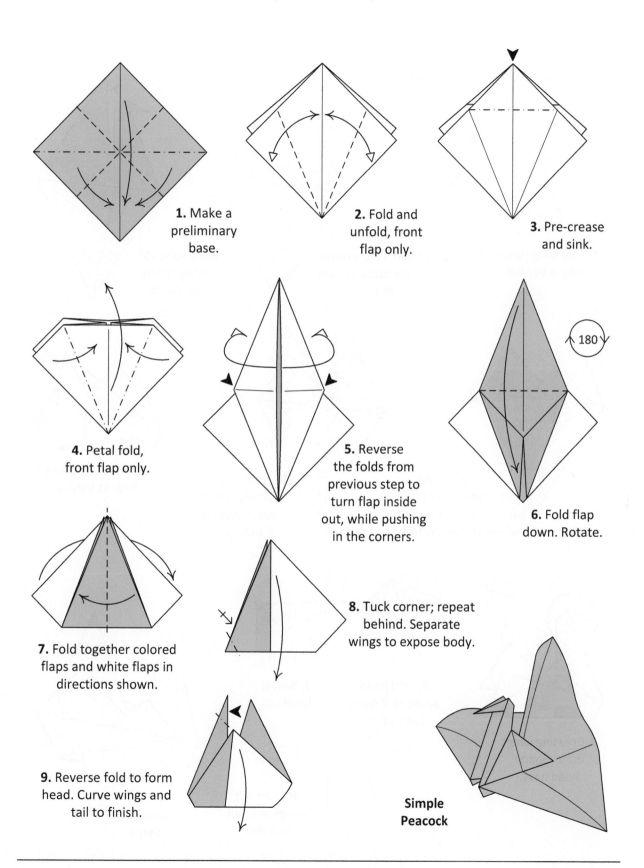

1. Make a preliminary base.

2. Fold and unfold, front flap only.

3. Pre-crease and sink.

4. Petal fold, front flap only.

5. Reverse the folds from previous step to turn flap inside out, while pushing in the corners.

6. Fold flap down. Rotate.

7. Fold together colored flaps and white flaps in directions shown.

8. Tuck corner; repeat behind. Separate wings to expose body.

9. Reverse fold to form head. Curve wings and tail to finish.

Simple Peacock

Simple Swan

(Created 2017)

Start with a completed Simple Peacock as on the precious page, flattened. Then continue as below.

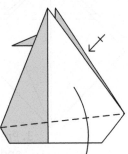

1. Fold wing down. Repeat behind.

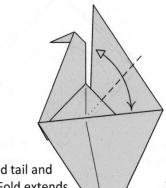

2. Fold tail and unfold. Fold extends inside body, shown dotted.

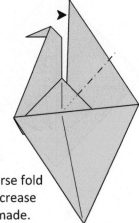

3. Reverse fold along crease just made.

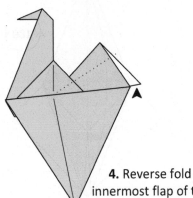

4. Reverse fold innermost flap of tail back up, approximately where shown dotted.

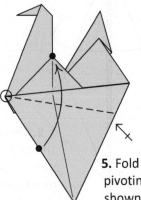

5. Fold wing up, pivoting where shown. Repeat behind.

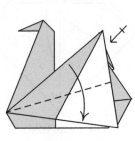

6. Fold wing down. Repeat behind.

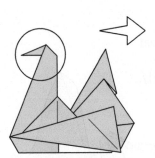

7. Creasing wings done. Shape head next.

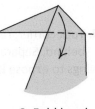

8. Fold head down as shown, both sides.

9. Swivel head up.

10. Fan out wings and tail to finish.

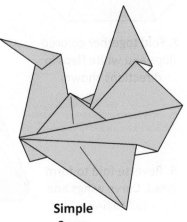

Simple Swan

Flower Strips

(Created 2017)

This is a fun, magical, color change design from a single strip of paper. It almost seems like the flowers have been stuck onto the strip, but, on the contrary, no glue or cutting is involved.

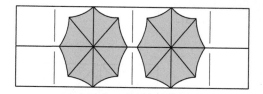

In the diagrams, we will illustrate how to fold a strip with two flowers. You can then extrapolate to any number of flowers you want using longer strips as per the chart below. You can also use wider strips and change the length proportionately.

Paper Size Chart

# of Flowers	Width	Length
2	2"	7"
3	2"	10"
4	2"	13"
n	2"	$(3n + 1)$"

1. For two flowers, start with a strip 2" by 7". Flower color side up, valley fold quarters as shown and unfold all.

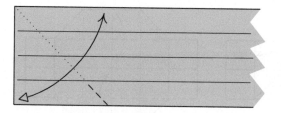

2. Working on the left side of the strip, pinch at the bottom of the diagonal and then turn over vertically.

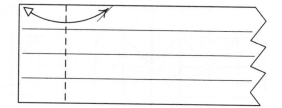

3. Fold edge to pinch point and unfold.

4. Fold edge to new crease.

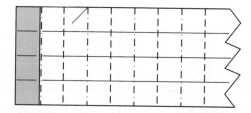

5. Fan fold the entire length of the paper by alternating mountain and valley folds. Then unfold all and turn over.

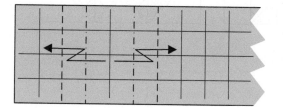

6. Using existing creases, make a box pleat.

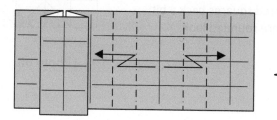

7. Fold the next box pleat.

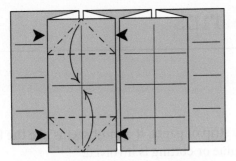

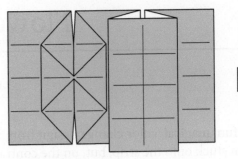

8. Squash using existing valley folds and forming new mountain folds.

9. Repeat previous step for the next box pleat.

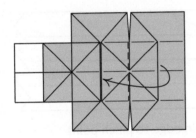

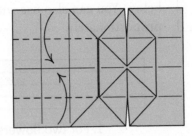

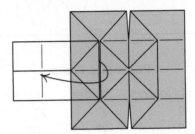

10. Turn flap to the right.

11. Fold along existing creases.

12. Turn flap back to the left.

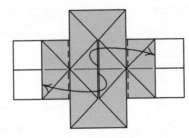

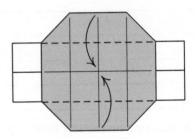

13. Repeat Steps 10–12 on the right flap, mirrored.

14. Turn the two central flaps away from each other.

15. Fold along existing creases.

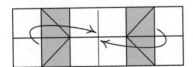

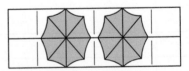

16. Return flaps back to the center.

17. Raise and shape the flowers by reinforcing the folds as shown.

18. Complete Flower Strip with two flowers.

Variation: Before performing Step 17, fold the corners of the colored squares back a bit for a more rounded look.

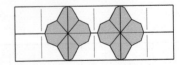

(Photo on p. 10.)

Hexagonal Flower

(Create 2017)

Start with a hexagon as shown on p. 9.

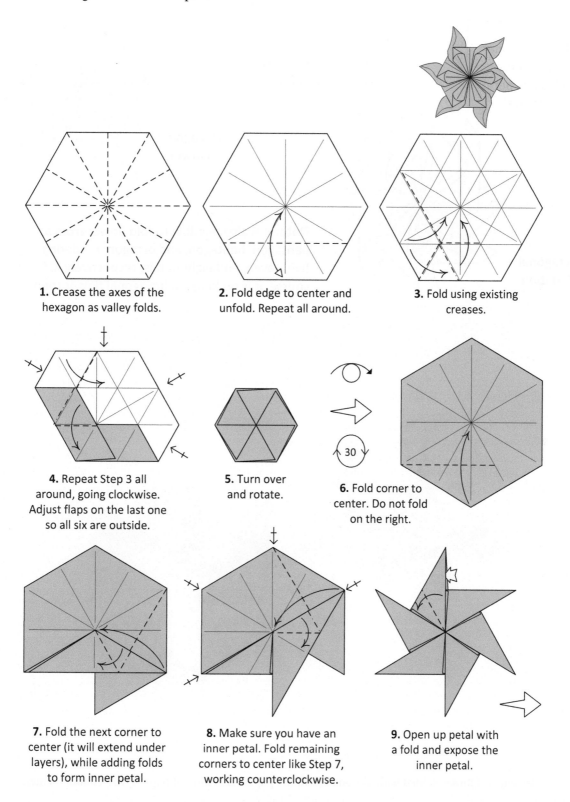

1. Crease the axes of the hexagon as valley folds.

2. Fold edge to center and unfold. Repeat all around.

3. Fold using existing creases.

4. Repeat Step 3 all around, going clockwise. Adjust flaps on the last one so all six are outside.

5. Turn over and rotate.

6. Fold corner to center. Do not fold on the right.

7. Fold the next corner to center (it will extend under layers), while adding folds to form inner petal.

8. Make sure you have an inner petal. Fold remaining corners to center like Step 7, working counterclockwise.

9. Open up petal with a fold and expose the inner petal.

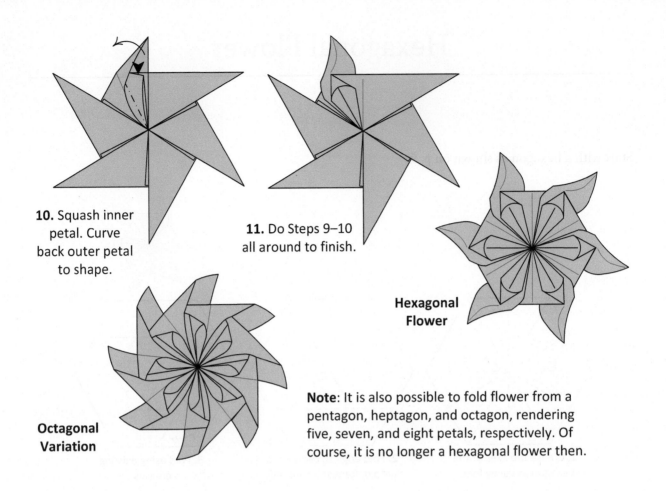

10. Squash inner petal. Curve back outer petal to shape.

11. Do Steps 9–10 all around to finish.

Hexagonal Flower

Octagonal Variation

Note: It is also possible to fold flower from a pentagon, heptagon, and octagon, rendering five, seven, and eight petals, respectively. Of course, it is no longer a hexagonal flower then.

Hexagonal Flowers folded with Corona Harmony paper. This kind of paper gives a darker center.

3 ◆ Four-Sink Base Flowers

(Created 2013)

Mum (p. 33) and Hollyhock (p. 35).

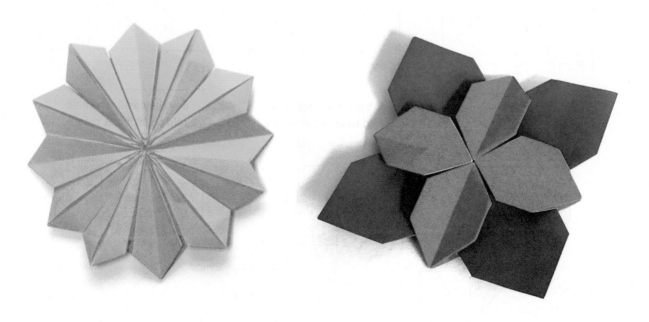

Marigold (p. 36) and Poinsettia with Leaves (p. 39).

Four-Sink Base

The Four-Sink Windmill Base, also known as the Four-Sink Base, is a very versatile base with endless possibilities. The base is essentially a Windmill Base with its four corners sunk, and hence the name. There are several ways to fold the base. I have presented three ways for different folding skill levels. Method 1 is for the novice, Method 2 is for people with more experience, and Method 3 is for people who are comfortable with collapses.

Method 1 (for Beginners)

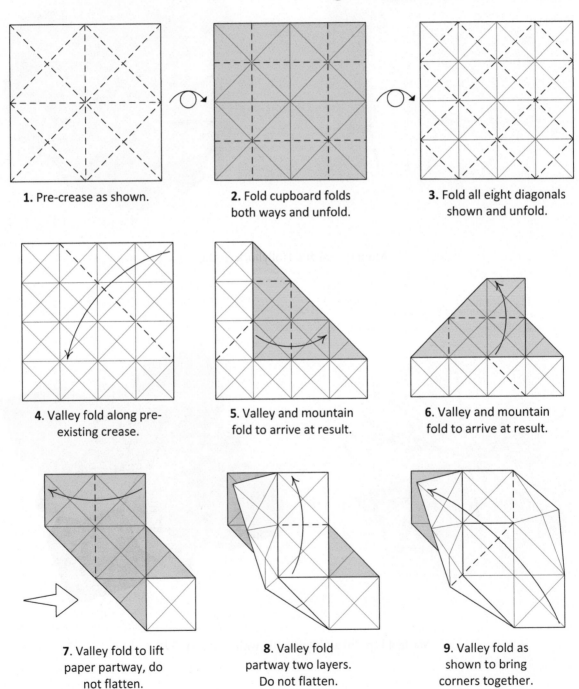

1. Pre-crease as shown.

2. Fold cupboard folds both ways and unfold.

3. Fold all eight diagonals shown and unfold.

4. Valley fold along pre-existing crease.

5. Valley and mountain fold to arrive at result.

6. Valley and mountain fold to arrive at result.

7. Valley fold to lift paper partway, do not flatten.

8. Valley fold partway two layers. Do not flatten.

9. Valley fold as shown to bring corners together.

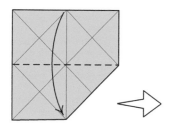

10. Valley fold top layers to arrive at result.

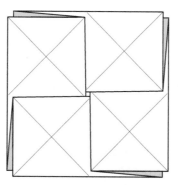

11. Turn over.

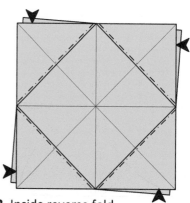

12. Inside reverse fold the four corners shown.

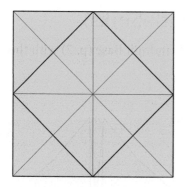

13. Turn over.

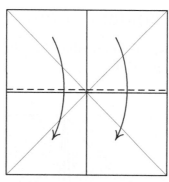

14. Valley fold left flap and then right flap.

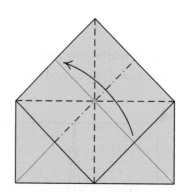

15. Collapse top flap like a Waterbomb Base.

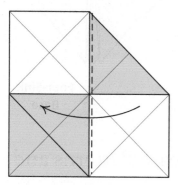

16. Valley fold top flap.

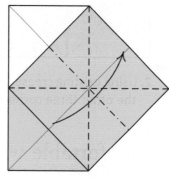

17. Collapse top flap like a Waterbomb Base.

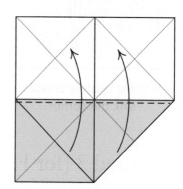

18. Valley fold left flap and then right flap.

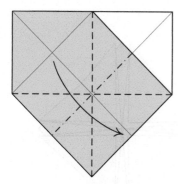

19. Collapse top flap like a Waterbomb Base.

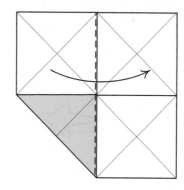

20. Valley fold top flap.

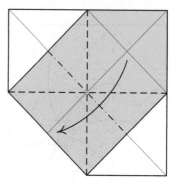

21. Collapse top flap like a Waterbomb Base.

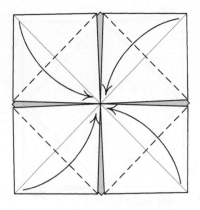

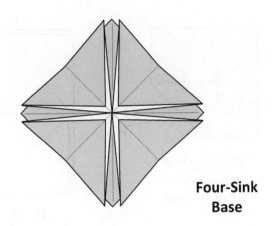

22. Bring corners
to center to
complete the base.

**Four-Sink
Base**

Method 2 (for Folders Comfortable with Sinks)

Start by folding the pre-creases of Steps 1–3 of Method 1 (p. 30). Make a Windmill Base (p. 7) with the
pre-creased paper, and follow the steps below.

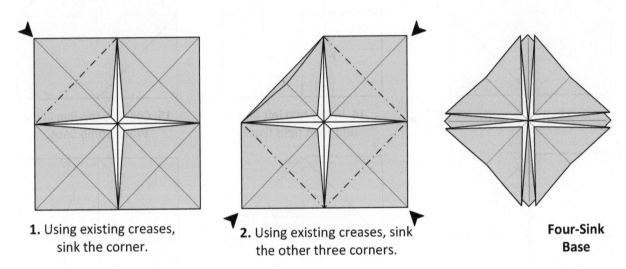

1. Using existing creases,
sink the corner.

2. Using existing creases, sink
the other three corners.

**Four-Sink
Base**

Method 3 (for Folders Comfortable with Crease Patterns and Collapses)

Start by folding the pre-creases of Steps 1–3 of Method 1 (p. 30).

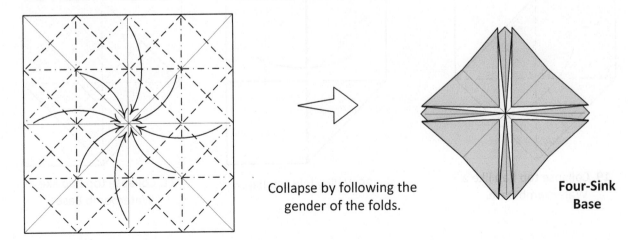

Collapse by following the
gender of the folds.

**Four-Sink
Base**

Mum

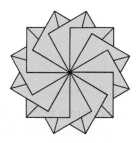

This flower is inspired by Hajime Komiya's beautiful creation, the Dahlia [Kom16]. The idea of obtaining twelve petals comes from his design, although the starting base and reference points are different, among other things.

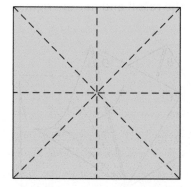

1. Start with 6" or larger square. Make the creases shown.

2. Fold in quarters both ways and unfold.

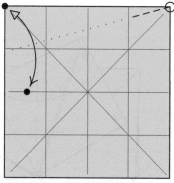

3. Bring corner to half fold, creasing only where shown.

4. Fold and unfold all the diagonals shown.

5. Make a Four-Sink Base as explained in the beginning of this chapter.

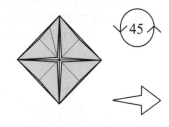

6. You now have a complete base with one extra crease. Rotate counterclockwise 45°.

For the next several steps we will focus on the part circled. Note that the extra crease made in Step 3 is on the left flap.

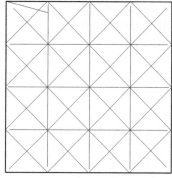

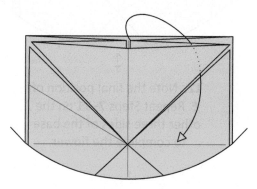

7. Pull middle flap out to the horizontal crease.

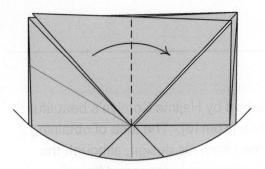

8. Turn left flap
to the right.

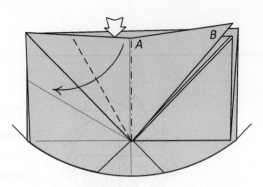

9. Bring the mountain fold to the reference crease made in Step 3. Flatten only the left side of the result and not the right.

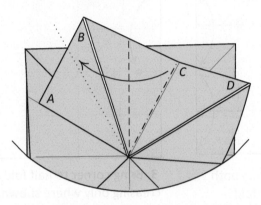

10. Note the final positions of *A* and *B*. Imagine the dotted line shown. Bring the mountain fold to the imaginary line. Flatten only the left side of the result and not the right side.

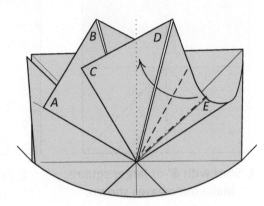

11. Note the final positions of *C* and *D*. Repeat Step 10 with the next flap. Note that the mountain fold is on an existing crease just like in Steps 9 and 10.

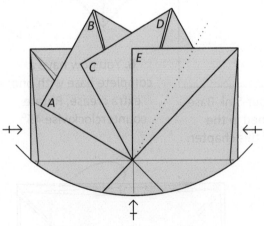

12. Note the final position of *E*. Repeat Steps 7–11 on the other three sides of the base to complete the flower.

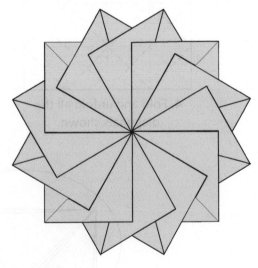

Mum

(Photo on p. 29.)

Hollyhock

This flower is only a minor variation of the previous flower, the Mum, with two additional finishing folds on each petal. However, it has a surprisingly different finished look.

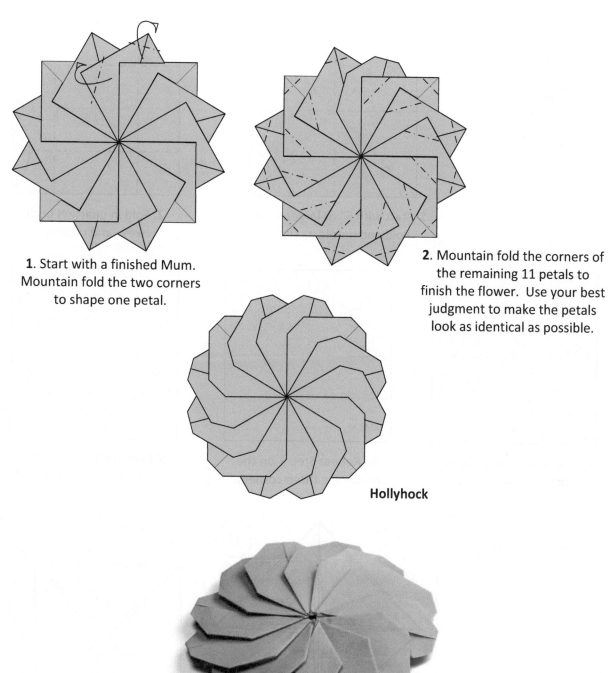

1. Start with a finished Mum. Mountain fold the two corners to shape one petal.

2. Mountain fold the corners of the remaining 11 petals to finish the flower. Use your best judgment to make the petals look as identical as possible.

Hollyhock

Hollyhock

Marigold

This Marigold design is inspired by Jorge Jaramillo's [Jar05] Marigold design and is a variation of his original version. Actually, we both inspired each other to arrive at our results.

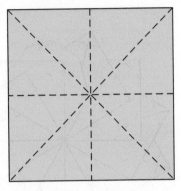

1. Start with 6" or larger square. Fold and unfold as shown.

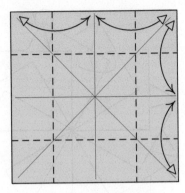

2. Fold in quarters both ways and unfold.

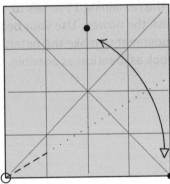

3. Bring corner to book fold, creasing only where shown.

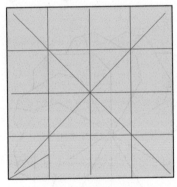

4. Repeat Step 3 on the remaining three corners.

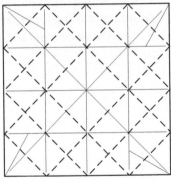

5. Fold and unfold all the diagonals shown.

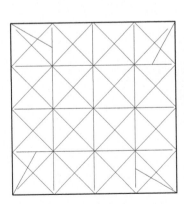

6. Make a Four-Sink Base as explained in the beginning of this chapter.

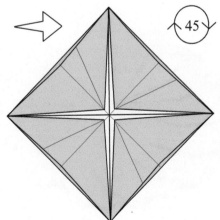

7. You now have a complete base with four extra creases. Rotate 45°.

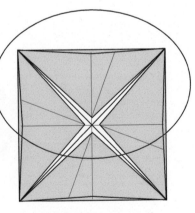

For the next several steps we will focus on the part circled.

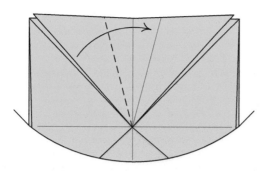

8. Fold left flap to
the existing crease.

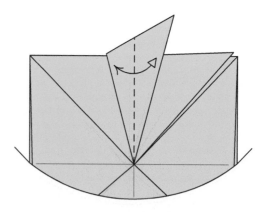

9. Valley fold flap
in half and unfold.

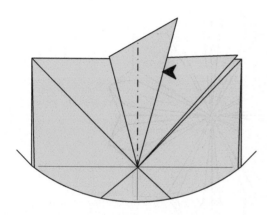

10. Squash flap using
crease just made.

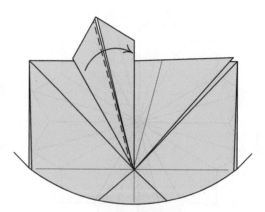

11. Fold the
petal in half.

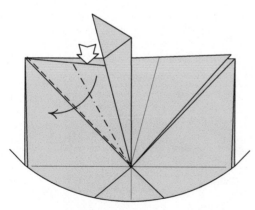

12. Pull paper out from behind
the petal, as far as it will go.
Flatten using creases shown.

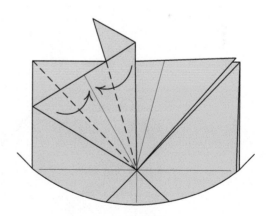

13. Valley fold the
two flaps as shown.

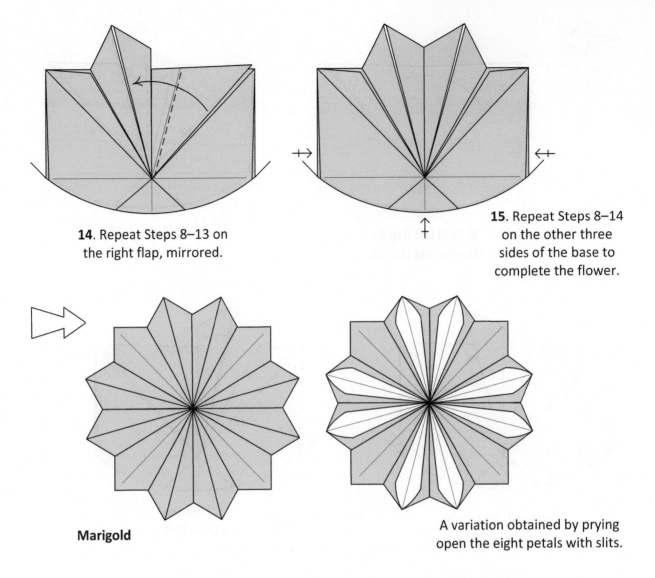

14. Repeat Steps 8–13 on the right flap, mirrored.

15. Repeat Steps 8–14 on the other three sides of the base to complete the flower.

Marigold

A variation obtained by prying open the eight petals with slits.

Marigold Variation. (Marigold photo on p. 29.)

Poinsettia with Leaves

(Created 2015)

Start with a completed Four-Sink Base (p. 30) and continue as below.

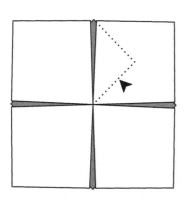

1. Reverse fold the hidden flap to bring it outside.

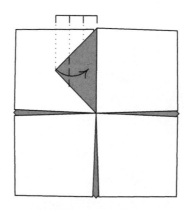

2. Fold about a third of the corner.

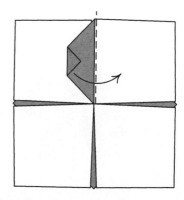

3. Turn top flap to the right.

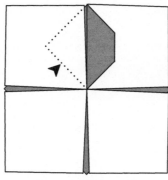

4. Repeat Steps 1–3 on the left, mirrored.

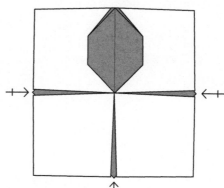

5. Repeat Steps 1–4 all around.

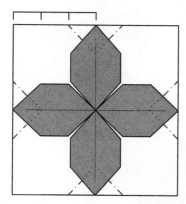

6. Shape the leaves by reaching between layers and folding back thirds. Folds extend behind petals.

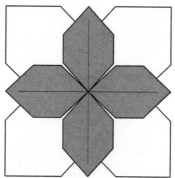

Poinsettia with Leaves
(folded with red *kami* and red-green duo)

(Photo on p. 29. Also in wreath on p. xvi.)

4 ◆ Kusumita and Derivatives

(Created 2013)

Kusumita Flowers and Leaves.

Kusumita derivatives: Butterflies, and Card Suits folded with *kami* and duo papers.

Modified Four-Sink Base

All the designs in this chapter start with the Modified Four-Sink Base. It is very much like the original Four-Sink Base, with the main difference being that it incorporates color change. The four front flaps and the square back of the modified base have colors from opposite sides of the paper. This enhances the beauty of the series of designs presented in this chapter. If you are not seeking color change, the designs may be folded from the original Four-Sink Base as well. I recommend using 7" *kami* paper.

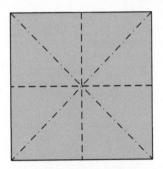

1. Fold and unfold both diagonals and book folds as shown.

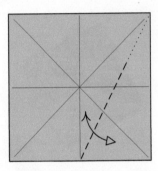

2. Fold from center of bottom edge to top right corner, not creasing all the way.

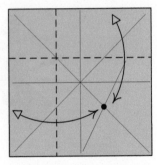

3. Fold and unfold edges to reference point just made.

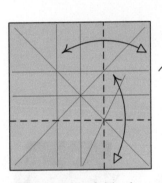

4. Fold and unfold edges to creases just made.

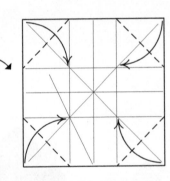

5. Fold all four corners in as shown.

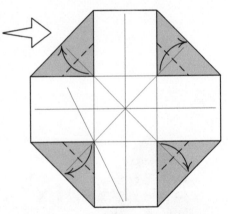

6. Fold all four corners out to edge.

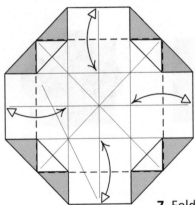

7. Fold and unfold the edges to creases made in Steps 3 and 4.

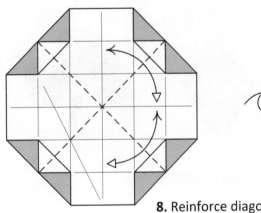

8. Reinforce diagonals as valley folds through all layers.

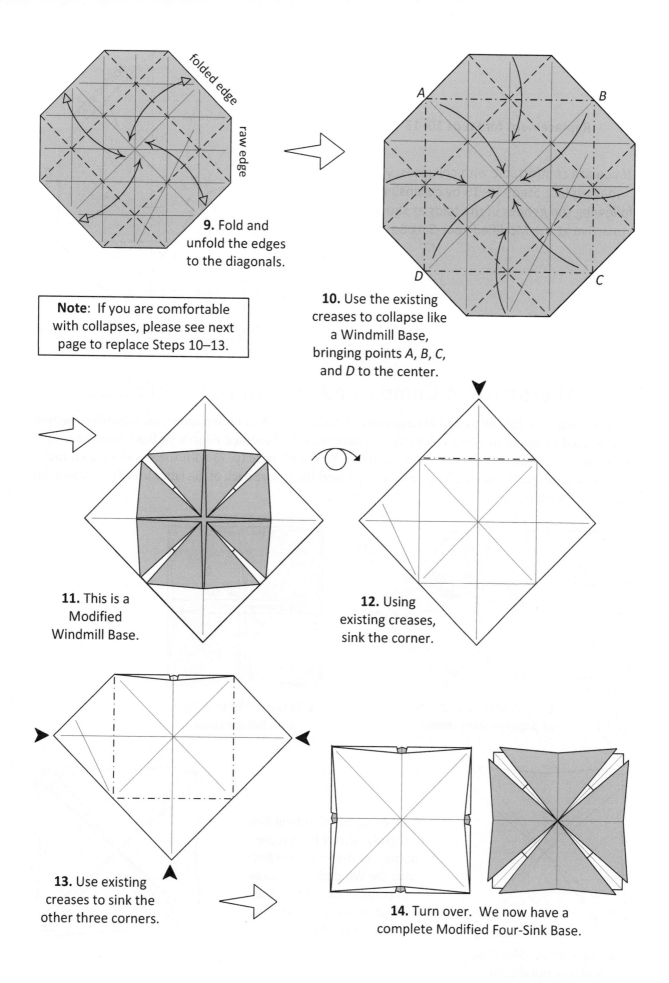

9. Fold and unfold the edges to the diagonals.

folded edge
raw edge

Note: If you are comfortable with collapses, please see next page to replace Steps 10–13.

10. Use the existing creases to collapse like a Windmill Base, bringing points *A*, *B*, *C*, and *D* to the center.

A   *B*
D   *C*

11. This is a Modified Windmill Base.

12. Using existing creases, sink the corner.

13. Use existing creases to sink the other three corners.

14. Turn over. We now have a complete Modified Four-Sink Base.

Replacement for Steps 10–13

If you are comfortable with collapses, you may replace Steps 10–13 by collapsing following the gender of the folds as shown to arrive at the Modified Four-Sink Base.

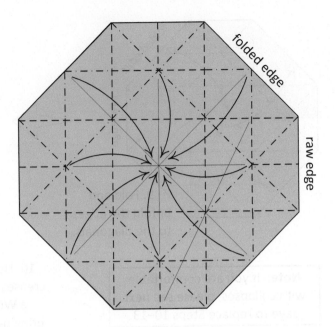

Alternative: A Composite Modified Four-Sink Base

Since duo paper is difficult to find in large sizes or in desired color combinations, the following method may be used to obtain the same color change effect as in the Modified Four-Sink Base. You can simply use *kami* of two different sizes, *a* and $\sqrt{2}a$, fold a Windmill Base (p. 7) with the smaller paper, fold a Four-Sink Base (p. 30) with the larger paper, and insert the square back of the latter into the former with the four flaps exposed.

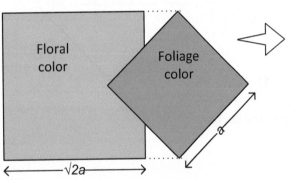

1. Start with two sheets of paper in sizes shown.

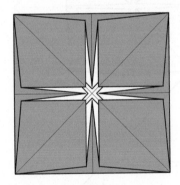

2. Make a Windmill Base with foliage color.

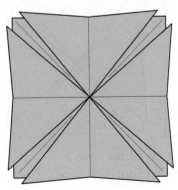

3. Make a Four-Sink Base with the floral color.

4. Open up the Windmill Base slightly. Insert the square bottom of the Four-Sink Base into the Windmill Base so as to completely hide it. The result should look like the figure on the right.

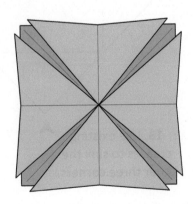

Modified Four-Sink Base

Kusumita

Kusumita is a Sanskrit word meaning "blossomed." The word is also used in the formal versions of Bengali (কুসুমিত), Hindi (कुसुमित), and other Indian languages derived from Sanskrit. Start with a Modified Four-Sink Base (p. 42) made from 7″ or larger paper.

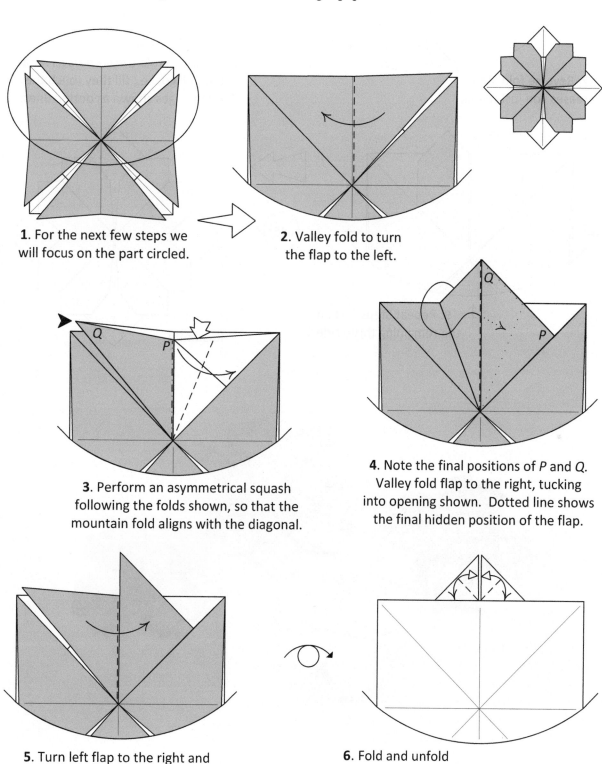

1. For the next few steps we will focus on the part circled.

2. Valley fold to turn the flap to the left.

3. Perform an asymmetrical squash following the folds shown, so that the mountain fold aligns with the diagonal.

4. Note the final positions of *P* and *Q*. Valley fold flap to the right, tucking into opening shown. Dotted line shows the final hidden position of the flap.

5. Turn left flap to the right and repeat Steps 3 and 4, mirrored.

6. Fold and unfold both tips.

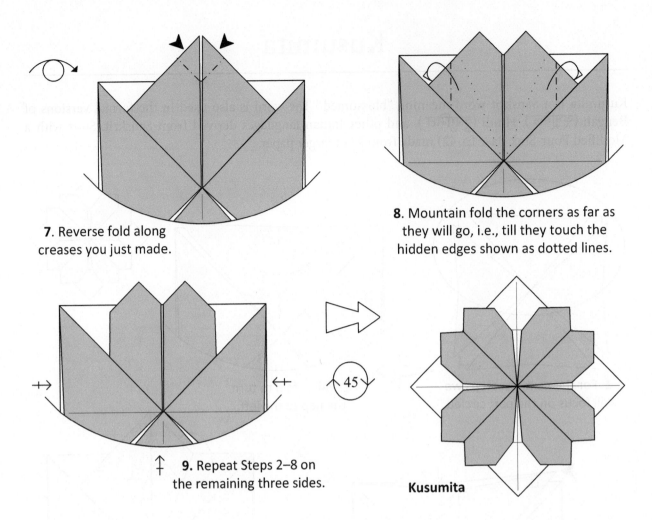

7. Reverse fold along creases you just made.

8. Mountain fold the corners as far as they will go, i.e., till they touch the hidden edges shown as dotted lines.

9. Repeat Steps 2–8 on the remaining three sides.

45

Kusumita

Kusumita in various colors. Folded in green, it looks like a Four Leaf Clover.

Kusumita Tiles

The Kusumita design has a large number of additional designs that can be derived from it. We will now identify some building blocks or *tiles* and see how to make dozens of derivatives. Each tile may be used in multiple designs. The purpose of this section is to organize the tiles in one place so that material is condensed and not repeated over and over again for each design.

The derivatives can be broadly categorized into flowers, leaves, butterflies, and playing card symbols. All of these begin with the Modified Four-Sink Base (p. 42). Each of the four flaps of the base can be finished into one of 12 or more distinct tiles. By mixing and matching the tiles, one can arrive at myriad interesting final results.

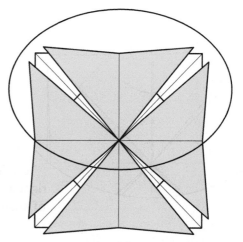

For each of the tiles, we will focus only on one flap of the Modified Four-Sink Base as circled.

Tile 1

Do Steps 1–7 of Kusumita (p. 45) to arrive at Tile 1.

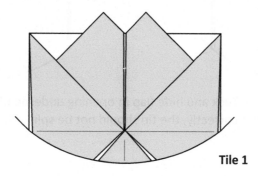

Tile 1

Tile 2

Do Steps 1–8 of Kusumita (p. 45) to arrive at Tile 2. This is the original Kusumita Petal.

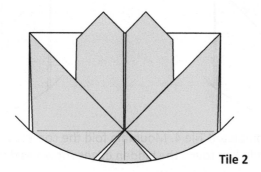

Tile 2

Tile 3

Do Steps 1–4 of Kusumita (p. 45) and then continue as below.

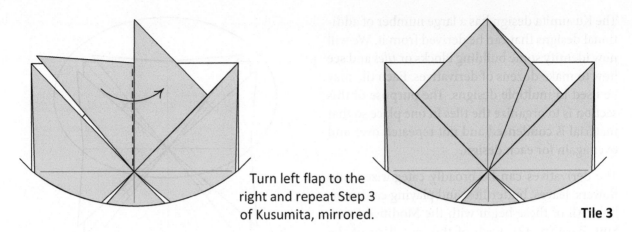

Turn left flap to the right and repeat Step 3 of Kusumita, mirrored.

Tile 3

Tile 4

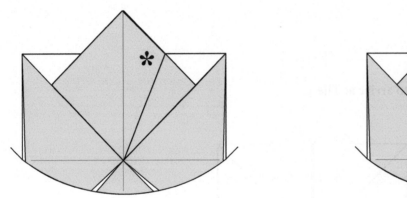

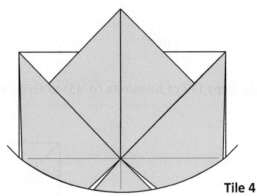

Start with Tile 3. Tuck and hide flap in opening underneath on the right. If you have tucked it correctly, the tip should not be split. Please double check.

Tile 4

Tile 5

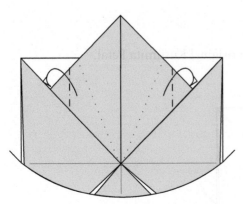

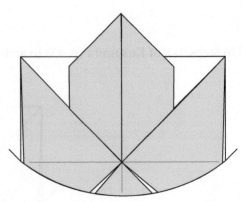

Start with a completed Tile 4. Mountain fold the corners as far as they will go, i.e., till they touch the hidden edges shown as dotted lines.

Tile 5

Tile 6

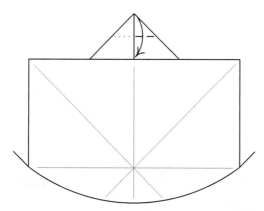

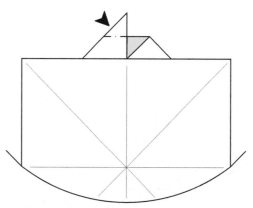

1. Start with a completed Tile 4 turned over. Valley fold right tip. The fold extends within the left flap as shown dotted.

2. Reverse fold left tip. The folded material should lie in front of the middle layer hidden inside.

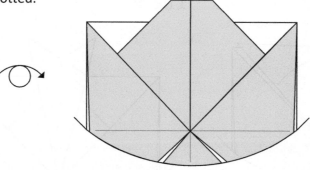

Tile 6

Tile 7

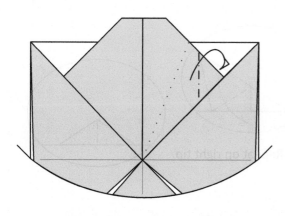

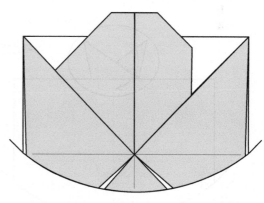

Tile 7

Start with a completed Tile 6. Mountain fold right corner as far as it will go, i.e., till it touches the hidden edge shown as dotted line.

Tile 8

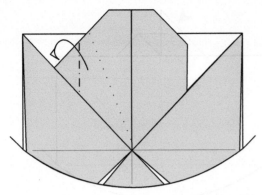

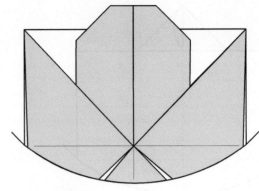

Tile 8

Start with Tile 7 and mountain fold
the left corner as far as it will go.

Tile 9

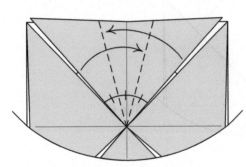

1. Trisect flap by trial and error.
Make creases sharp only when
the three sections are equal.

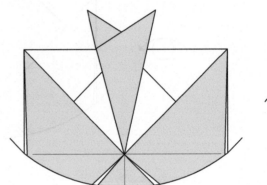

2. Turn over.

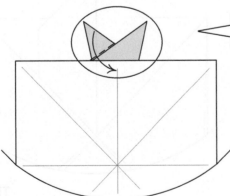

3. Valley fold along
edge and tuck tip
behind white square.

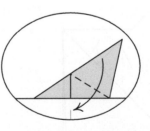

4. Repeat on right tip.

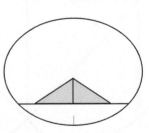

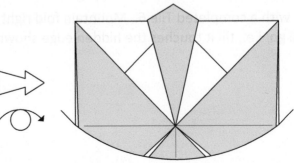

Tile 9

Tile 10

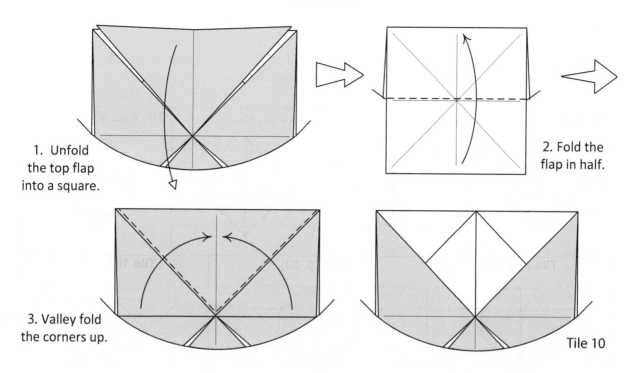

1. Unfold the top flap into a square.

2. Fold the flap in half.

3. Valley fold the corners up.

Tile 10

A Summary of the Tiles

This is a quick reference table of all the tiles we discussed so far.

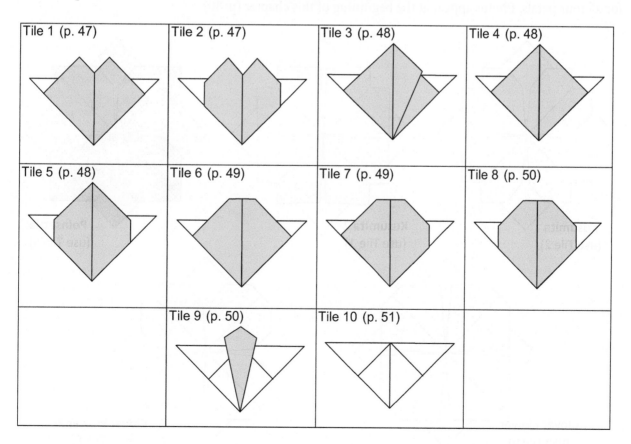

Tile 1 (p. 47)	Tile 2 (p. 47)	Tile 3 (p. 48)	Tile 4 (p. 48)
Tile 5 (p. 48)	Tile 6 (p. 49)	Tile 7 (p. 49)	Tile 8 (p. 50)
	Tile 9 (p. 50)	Tile 10 (p. 51)	

Flowers

Seven of the tiles discussed in the previous section (p. 47) will be used to make several different flowers, including the Kusumita. The tiles are listed in the following chart:

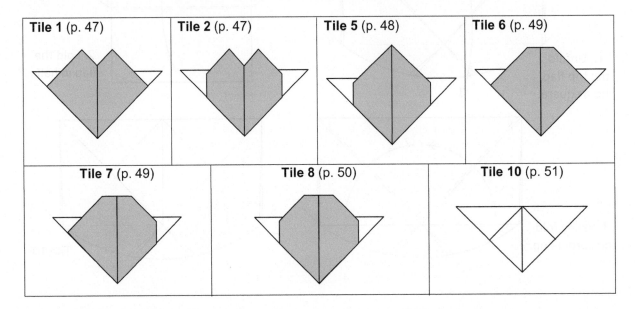

| **Tile 1** (p. 47) | **Tile 2** (p. 47) | **Tile 5** (p. 48) | **Tile 6** (p. 49) |

| **Tile 7** (p. 49) | **Tile 8** (p. 50) | **Tile 10** (p. 51) |

For each flower, start with a Modified Four-Sink Base (p. 42) and fold each flap of the base as indicated below to arrive at the various flowers, including the Kusumita. These flowers use the same kind of tiles for all four petals. Photos appear at the beginning of this chapter (p. 40).

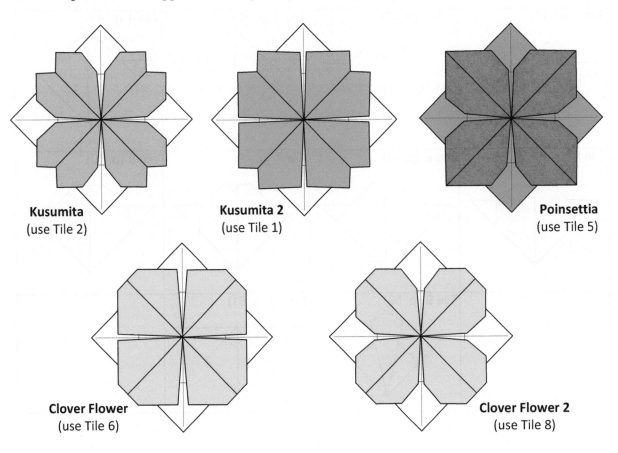

Kusumita
(use Tile 2)

Kusumita 2
(use Tile 1)

Poinsettia
(use Tile 5)

Clover Flower
(use Tile 6)

Clover Flower 2
(use Tile 8)

For Violas we will mix up tiles as below, making sure to use the blank Tile 10 for the top flap.

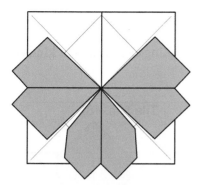

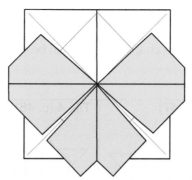

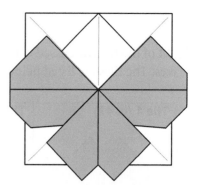

Viola 1
(use Tiles 1, 2, and 10)

Viola 2
(use Tiles 1, 8, and 10)

Viola 3
(use Tiles 1, 7, and 10)

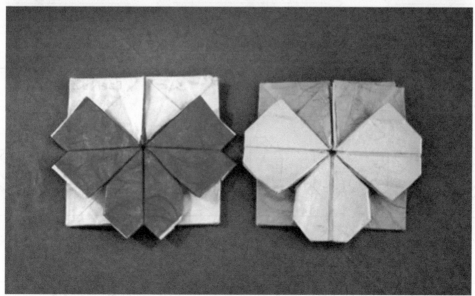

Top: Clover Flower and Kusumita 2. Bottom: Violas. More flower photos on p. 40.

Leaves

Eight of the tiles discussed earlier in this chapter (p. 47) will be used for making eight different kinds of leaves. The tiles are listed below:

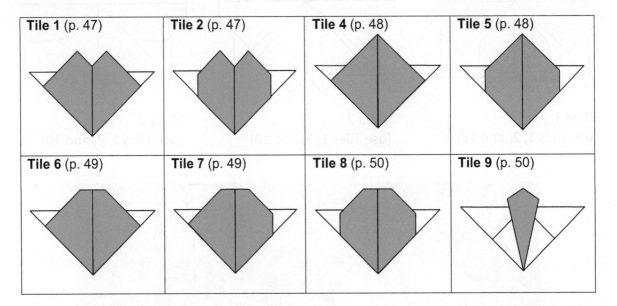

| Tile 1 (p. 47) | Tile 2 (p. 47) | Tile 4 (p. 48) | Tile 5 (p. 48) |
| Tile 6 (p. 49) | Tile 7 (p. 49) | Tile 8 (p. 50) | Tile 9 (p. 50) |

For each of the following leaves, start with a Modified Four-Sink Base (p. 42) and finish the bottom flap like Tile 9 to make the stalk. Finish the other three flaps of the base with the same kind of tile as indicated below:

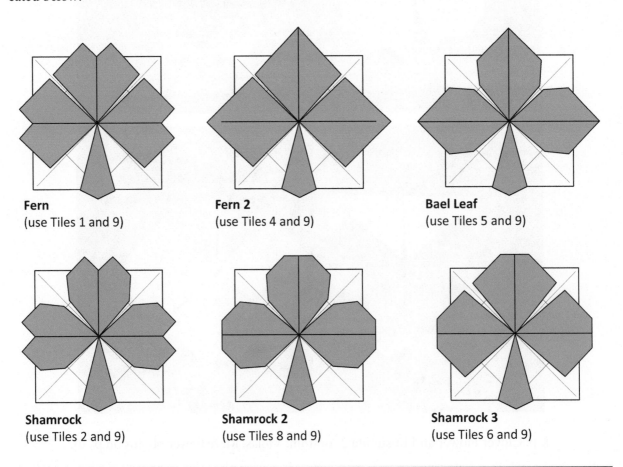

Fern
(use Tiles 1 and 9)

Fern 2
(use Tiles 4 and 9)

Bael Leaf
(use Tiles 5 and 9)

Shamrock
(use Tiles 2 and 9)

Shamrock 2
(use Tiles 8 and 9)

Shamrock 3
(use Tiles 6 and 9)

For the Four Leaf Clover and the Philodendron leaves, fold the flaps of a Modified Four-Sink Base (p. 42) as follows:

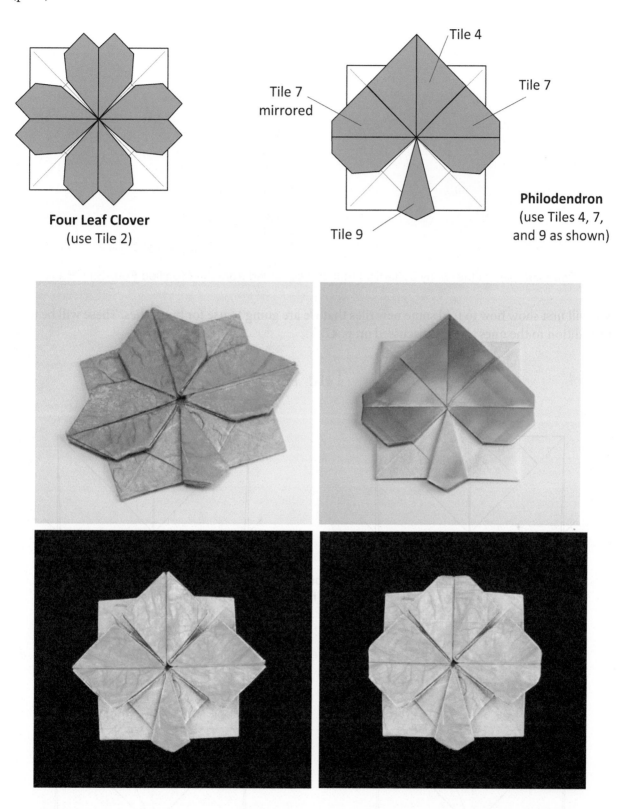

Four Leaf Clover
(use Tile 2)

Tile 4

Tile 7
mirrored

Tile 7

Tile 9

Philodendron
(use Tiles 4, 7,
and 9 as shown)

Top: Bael Leaf and Philodendron. Bottom: Fern 2 and Shamrock 3. More leaf photos on p. 40.
(Note that the Bael Leaf is considered to be sacred in Hinduism and is essential for many rituals.)

 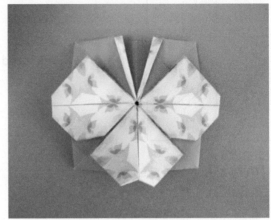

Two examples of the many butterflies that can be folded from the Modified Four-Sink Base

We will first show how to fold some new tiles that we are going to use for butterflies. These will be used in addition to the ones already discussed on p. 47.

Tile 11

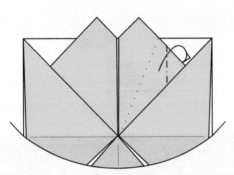 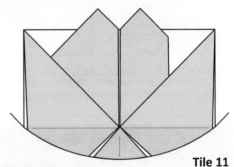

Tile 11

Start with a completed Tile 1. Mountain fold the right corner as far as it will go, i.e., till it touches the hidden edge shown as dotted line.

Tile 12

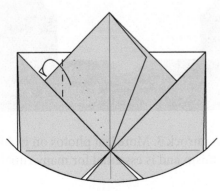 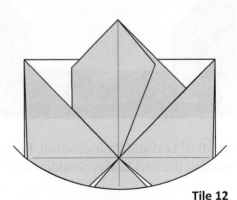

Tile 12

Start with a completed Tile 3. Mountain fold the left corner as far as it will go, i.e., till it touches the hidden edge shown as dotted line.

Tile 13

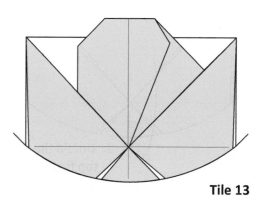

Tile 13

Start with a completed Tile 12. Mountain fold the top tip.

Tile 14

Do Steps 1–5 of Kusumita (p. 45) and continue with the following steps.

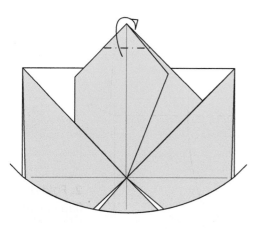

1. Fold and unfold right tip.

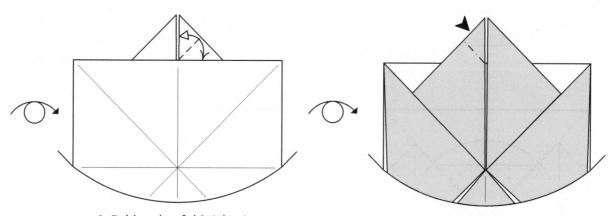

2. Reverse fold along crease you just made.

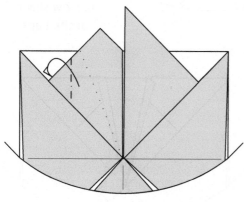

3. Mountain fold the left corner as far as it will go, i.e., till it touches the hidden edge shown as dotted line.

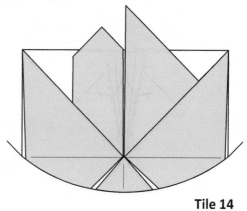

Tile 14

Tile 15

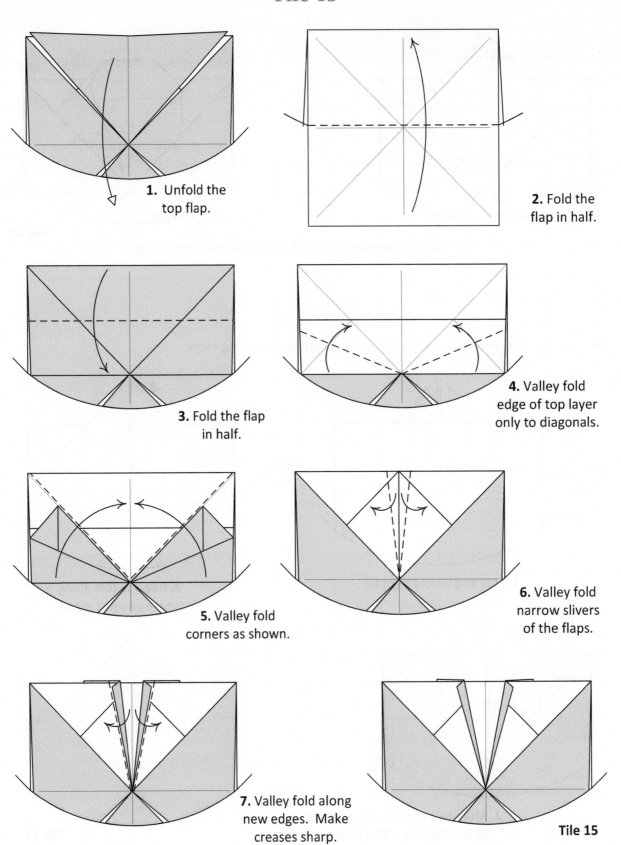

1. Unfold the top flap.

2. Fold the flap in half.

3. Fold the flap in half.

4. Valley fold edge of top layer only to diagonals.

5. Valley fold corners as shown.

6. Valley fold narrow slivers of the flaps.

7. Valley fold along new edges. Make creases sharp.

Tile 15

This tile has some very close variations, which are named 16 A, 16 B, 16 C, and 16 D.

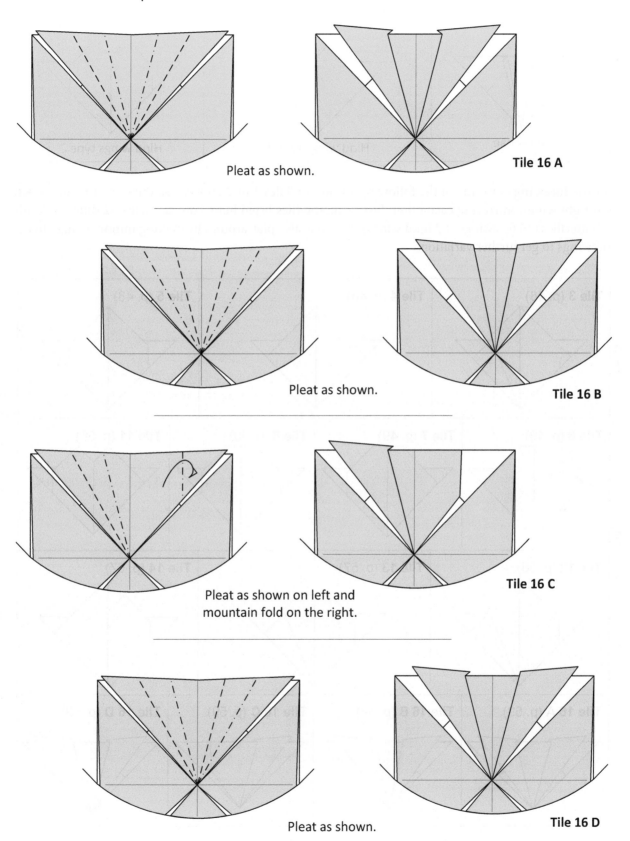

Pleat as shown.

Tile 16 A

Pleat as shown.

Tile 16 B

Pleat as shown on left and
mountain fold on the right.

Tile 16 C

Pleat as shown.

Tile 16 D

You can now build your own butterflies working on the flaps of a Modified Four-Sink Base (p. 42). For the **antennae** and **hind wings** of a butterfly, use the following tiles:

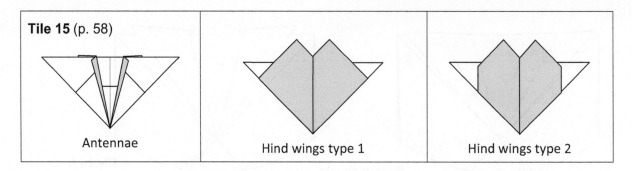

For the **forewings,** use any of the following 14 tiles or Tiles 1 or 2 above. Use the same tile for the left and right wings, mirroring each other. Just using the tiles listed here, you can make 32 different kinds of butterflies (16 forewings × 2 hind wings). You may also play around by making minor changes to the forewings to get further variations.

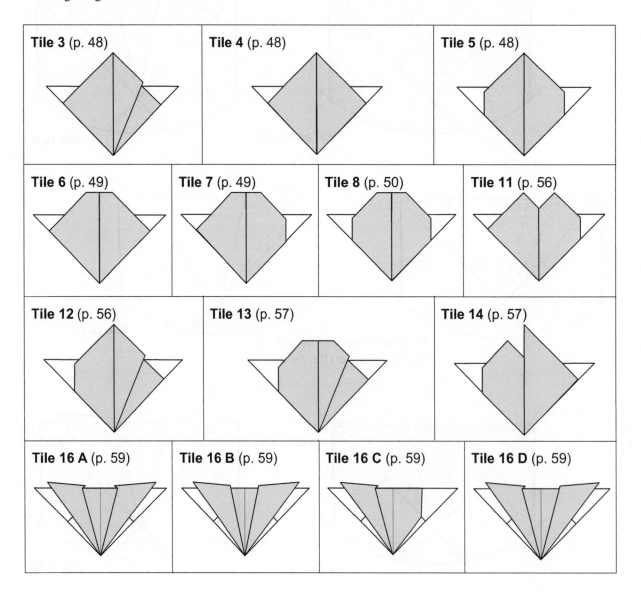

Below are shown 12 of the 32 different butterflies that you can potentially make. For each butterfly start with a Modified Four-Sink Base (p. 42), and finish each flap following the directions given on the previous page, i.e., use Tile 15 for the antennae, Tiles 1 or 2 for the hind wings, and any of Tiles 1–8, 11–14, or 16 for the forewings. You can even try your own variations on the flaps.

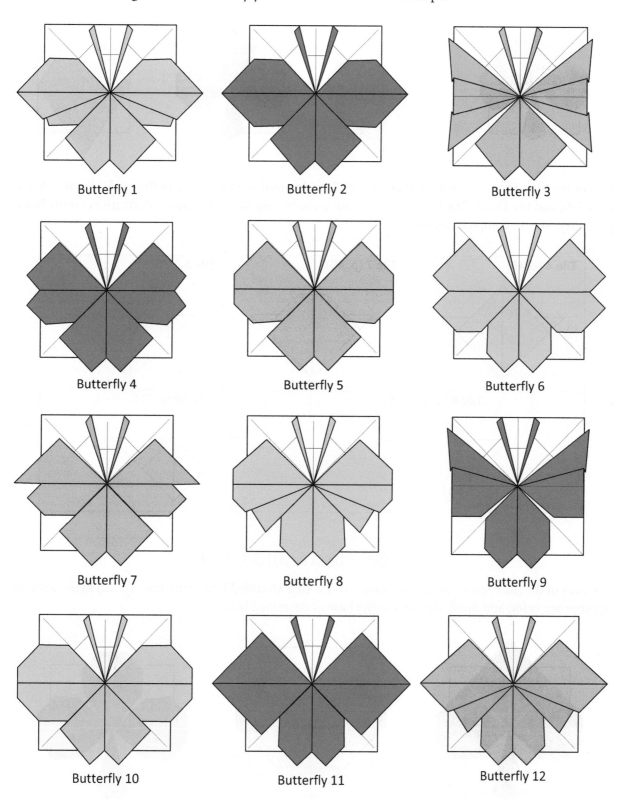

| Butterfly 1 | Butterfly 2 | Butterfly 3 |

| Butterfly 4 | Butterfly 5 | Butterfly 6 |

| Butterfly 7 | Butterfly 8 | Butterfly 9 |

| Butterfly 10 | Butterfly 11 | Butterfly 12 |

See pages 41 and 56 for photos of finished butterflies.

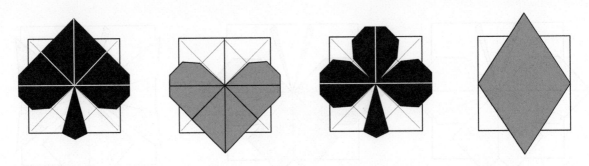

Five of the tiles discussed in the section on p. 47 can be used to make three of the card suits: the Spade, the Club, and the Heart. The Diamond does not quite belong to the Kusumita derivatives family but is presented here to complete the suit set.

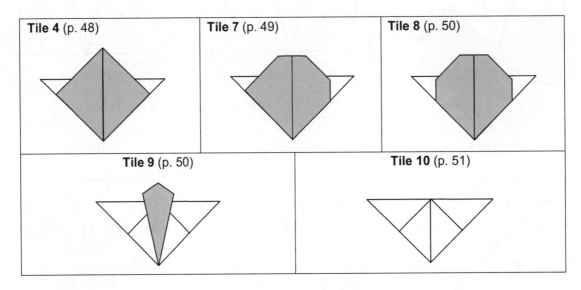

Tile 4 (p. 48)	Tile 7 (p. 49)	Tile 8 (p. 50)
Tile 9 (p. 50)		Tile 10 (p. 51)

Spade, Heart, and Club

For each of the suits Spade, Heart, and Club, start with a Modified Four-Sink Base (p. 42) made with the appropriate color, and finish the flaps of the base as described below:

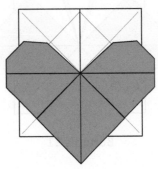

Spade (Top: Tile 4;
Left: Tile 7 mirrored;
Right: Tile 7; Bottom: Tile 9)

Heart (Top: Tile 10;
Left: Tile 7; Right: Tile 7
mirrored; Bottom: Tile 4)

Club
(Top, left, and right: Tile 8;
Bottom: Tile 9)

Diamond

This is the only design in the card suit set that does not start with a Modified Four-Sink Base. Start with the same size paper as the other members of the set.

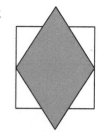

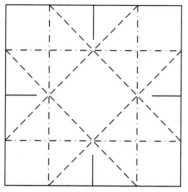

1. Pre-crease as shown the mountain and valley folds.

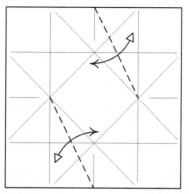

2. Make the two creases shown.

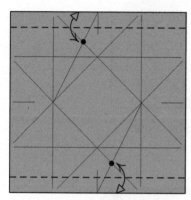

3. Fold and unfold edges to reference points just made.

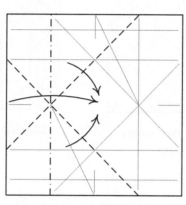

4. Collapse the left side using the existing valley and mountain creases.

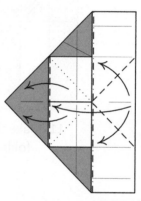

5. Valley fold white flaps to the left while repeating Step 4 on the right.

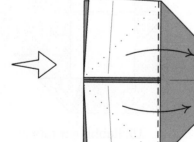

6. Valley fold flaps to the right while squashing the next layer getting pulled from behind.

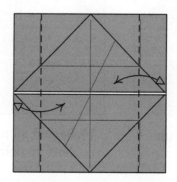

7. Fold and unfold through all layers using existing creases as reference.

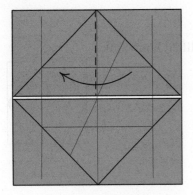

8. Fold top right flap to the left.

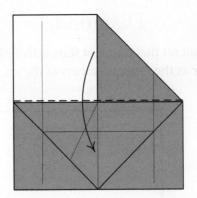

9. Turn top flap down.

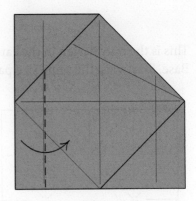

10. Fold front layer at existing crease.

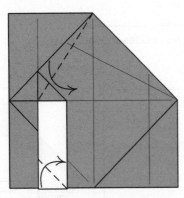

11. Make the two valley folds shown.

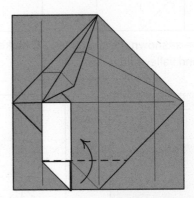

12. Valley fold as shown.

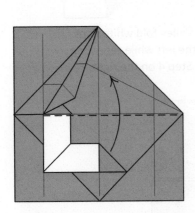

13. Turn flap back up.

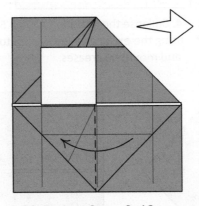

14. Repeat Steps 8–13 on bottom right flap.

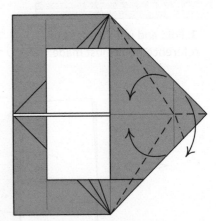

15. Rabbit ear fold at reference point shown.

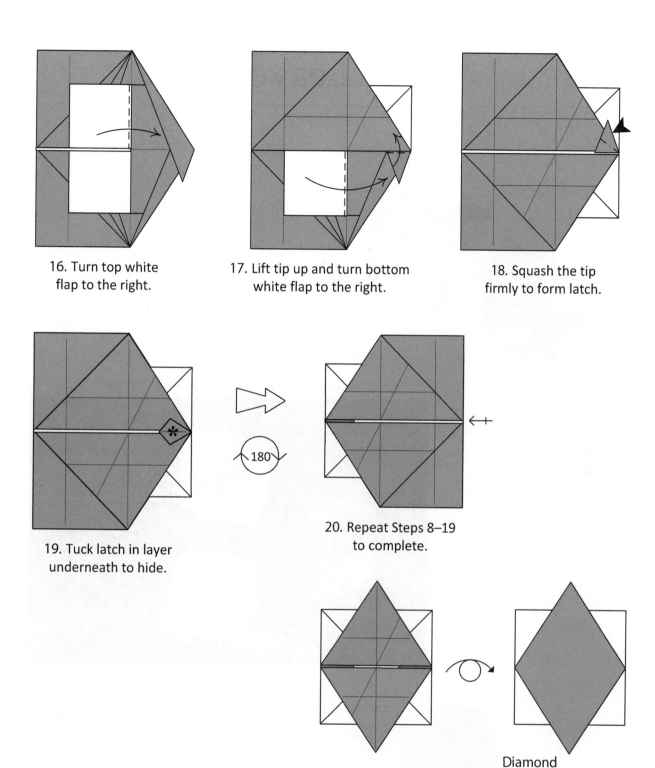

16. Turn top white flap to the right.

17. Lift tip up and turn bottom white flap to the right.

18. Squash the tip firmly to form latch.

19. Tuck latch in layer underneath to hide.

180

20. Repeat Steps 8–19 to complete.

Diamond

(Photos of Card Suits are on p. 41.)

Note that this diamond is almost the same as the Ace of Diamond on p. 13, but the white back on which the diamond rests is smaller in this case. This way the diamond juts out of the back to match the rest of the card suits.

♠ ♥ ♦ ♣ I dedicate my Card Suits to my octogenarian father, Shilananda Adhikari, who can still play a sharp game of Bridge. ♠ ♥ ♦ ♣

5 ◆ Leaves

Top: Tropical Leaves (p. 69). Middle: Elm Leaves (p. 71). Bottom: Aspen Leaves (p. 74).

Maple Leaves

(Created 2014 by David Donahue)

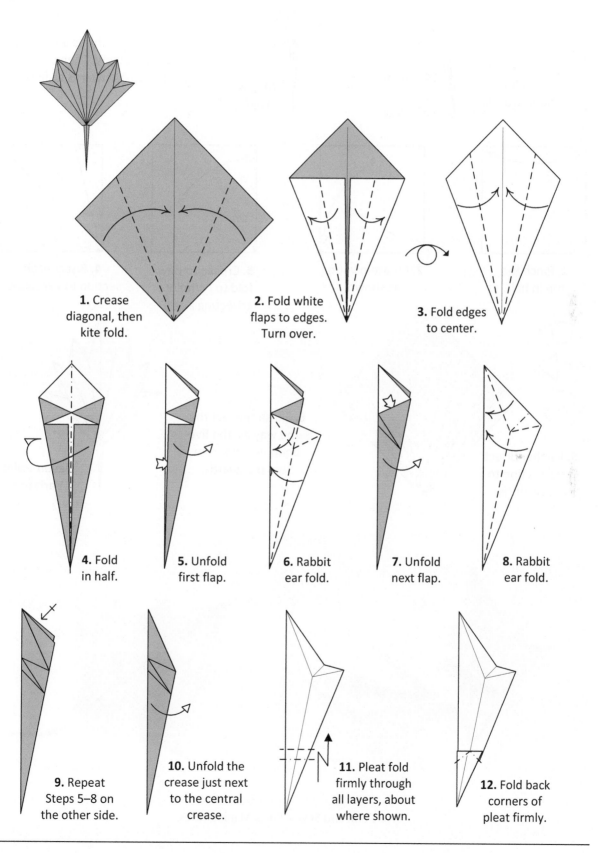

1. Crease diagonal, then kite fold.

2. Fold white flaps to edges. Turn over.

3. Fold edges to center.

4. Fold in half.

5. Unfold first flap.

6. Rabbit ear fold.

7. Unfold next flap.

8. Rabbit ear fold.

9. Repeat Steps 5–8 on the other side.

10. Unfold the crease just next to the central crease.

11. Pleat fold firmly through all layers, about where shown.

12. Fold back corners of pleat firmly.

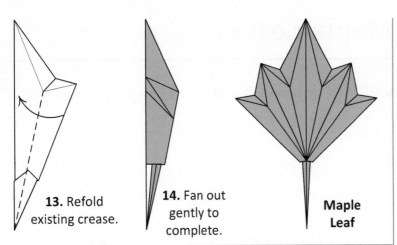

13. Refold existing crease.

14. Fan out gently to complete.

Maple Leaf

Author's note:
Using the same principles as the Maple Leaf, you can also fold a seven-point variation. In the first few steps, instead of folding halves and quarters of the angles on each side of the diagonal, fold thirds and sixths as shown below. Then continue from Step 6 onwards to finish.

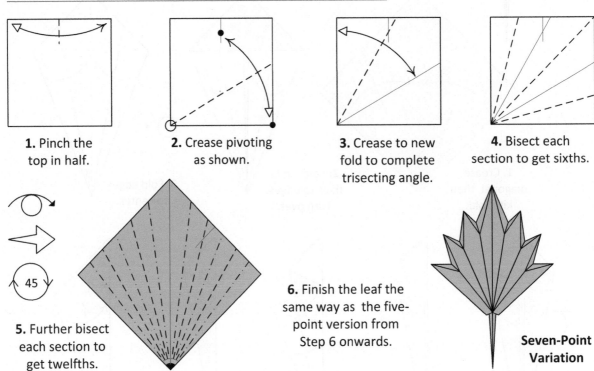

1. Pinch the top in half.

2. Crease pivoting as shown.

3. Crease to new fold to complete trisecting angle.

4. Bisect each section to get sixths.

5. Further bisect each section to get twelfths.

45

6. Finish the leaf the same way as the five-point version from Step 6 onwards.

Seven-Point Variation

Five-Point and Seven-Point Maple Leaves.

Maple Leaves

Tropical Leaf

(Created 2018)

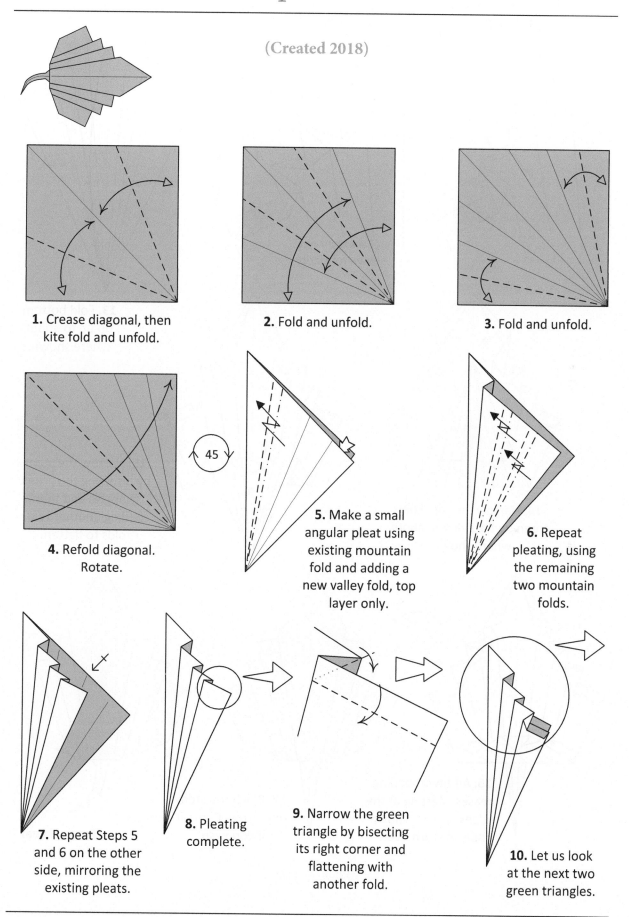

1. Crease diagonal, then kite fold and unfold.

2. Fold and unfold.

3. Fold and unfold.

4. Refold diagonal. Rotate.

45

5. Make a small angular pleat using existing mountain fold and adding a new valley fold, top layer only.

6. Repeat pleating, using the remaining two mountain folds.

7. Repeat Steps 5 and 6 on the other side, mirroring the existing pleats.

8. Pleating complete.

9. Narrow the green triangle by bisecting its right corner and flattening with another fold.

10. Let us look at the next two green triangles.

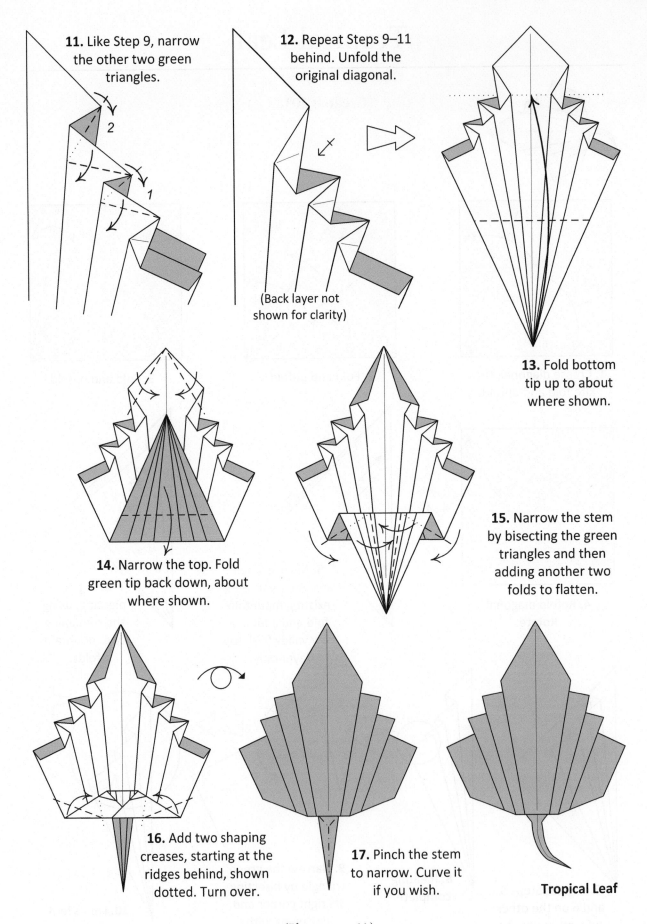

11. Like Step 9, narrow the other two green triangles.

12. Repeat Steps 9–11 behind. Unfold the original diagonal.

(Back layer not shown for clarity)

13. Fold bottom tip up to about where shown.

14. Narrow the top. Fold green tip back down, about where shown.

15. Narrow the stem by bisecting the green triangles and then adding another two folds to flatten.

16. Add two shaping creases, starting at the ridges behind, shown dotted. Turn over.

17. Pinch the stem to narrow. Curve it if you wish.

Tropical Leaf

(Photo on p. 66.)

Elm Leaf

(Created 2017)

Using a 6" square is recommended.

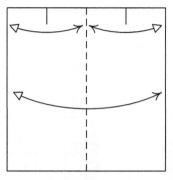

1. Fold in half and unfold. Pinch quarters.

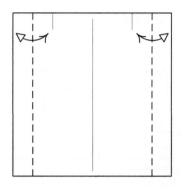

2. Using the pinch marks, fold the eighths. Unfold.

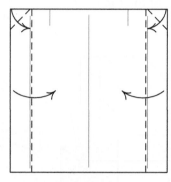

3. Fold the two corners, then refold existing creases.

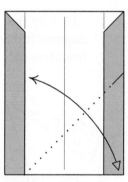

4. Make a small pinch on the right as shown.

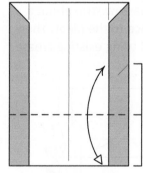

5. Fold to where pinch meets edge. Unfold.

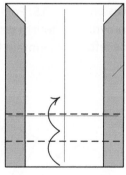

6. Fold twice as shown.

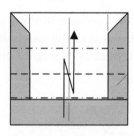

7. Pleat three more equal sections. Unfold to Step 4.

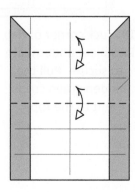

8. We now have six sections, five equal and the topmost one larger. Change the two mountain folds to valley folds. Turn over.

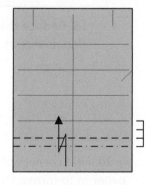

9. Make a pleat with the existing mountain fold and a new valley fold, which is close to but no more than a third of the section.

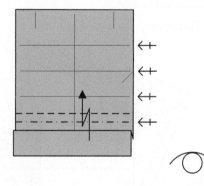

10. Similarly, repeat pleating the next four mountain folds, by adding new valley folds. Pleats should be about the same size and not overlap. Turn over.

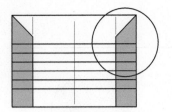

11. Pleating complete. We will now form the spikes of the leaf, starting at the top right.

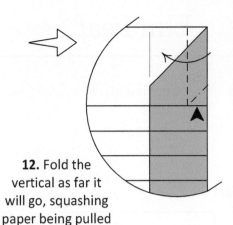

12. Fold the vertical as far it will go, squashing paper being pulled from behind.

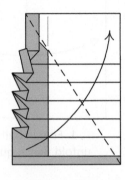

13. Work on the next pleat, reducing projection by half or more.

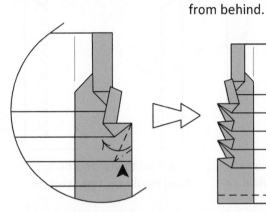

14. Continue forming spikes on the rest of the pleats as shown. Repeat from Step 12 on the left.

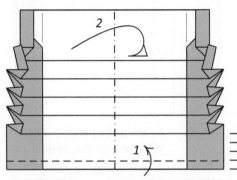

15. Fold a narrow strip about a fourth distance to the pleat. Then mountain fold using existing crease.

16. Fold the diagonal firmly, front flap only.

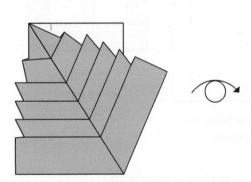

17. Turn over.

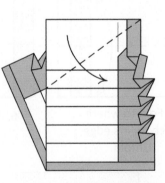

18. Reduce the flap as shown.

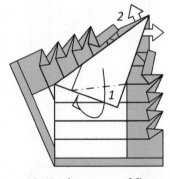

19. Tuck corner of flap into pleat underneath to secure. Pull the hidden green strips out to the top.

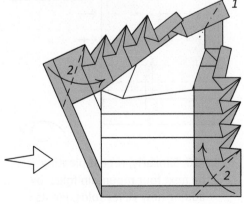

20. Fold corners of pulled paper in to form a sharp, neat tip. Fold bottom corners and tuck under pleats.

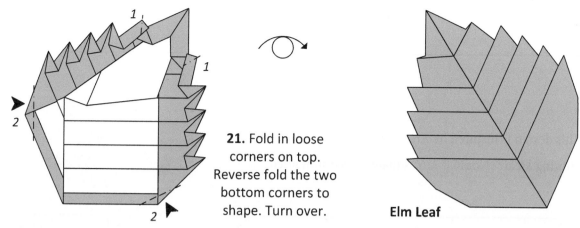

21. Fold in loose corners on top. Reverse fold the two bottom corners to shape. Turn over.

Elm Leaf

Variations

By starting with rectangles of various proportions, you can have many different shapes of leaves. You can also change the number and width of the pleats to obtain variations. Diagrams are not provided for the variations; rather it is left for you to experiment and discover.

Elm Leaf Variations. (Elm Leaf photos are on p. 66 and also in wreath on p. xvi.)

Aspen Leaf

(Created 2018)

This design is inspired by Atsuko Kawate's
Shining Heart [Noa17]. A 6″ square is recommended.

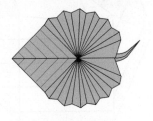

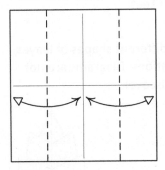

1. Crease both halves and the quarters shown.

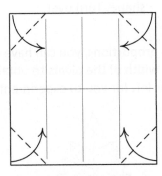

2. Fold the four corners.

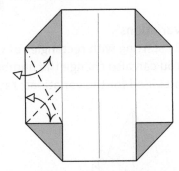

3. Crease the two diagonals shown. It is easier to mountain fold the longer one.

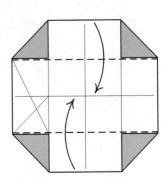

4. Fold and turn over.

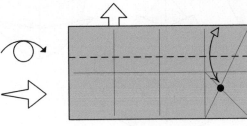

5. Fold edge to point while sliding out back flap. Unfold.

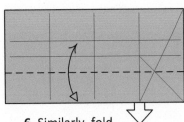

6. Similarly, fold bottom edge to new crease. Unfold.

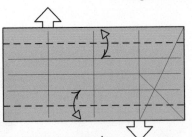

7. Add two more creases to complete folding sixths. Turn over.

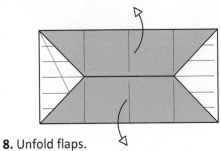

8. Unfold flaps.

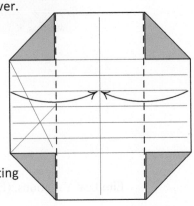

9. Fold existing creases.

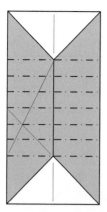

10. Reinforce horizontal creases as mountain folds.

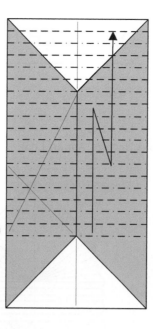

11. Accordion fold by using existing mountain folds and adding new valley folds, then continuing to the top section.

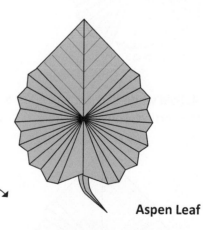

12. Unfold and flatten creases of top section. Make two 45° pleats and tuck in pockets. The accordion folds will get securely pinched at the center.

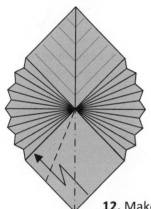

12. Make an angular pleat as shown.

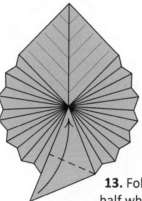

13. Fold triangle in half while squashing paper being pulled from behind.

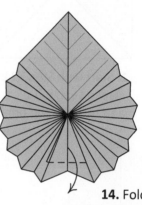

14. Fold flap down horizontally as far as it will go.

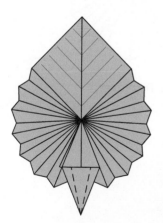

15. Narrow the stem and squash paper being pulled. Left side will get thick.

16. Curve stem to taste. Turn over.

Aspen Leaf

Note: If using paper smaller than a 6" square, make fewer accordion pleats.

(Photo on p. 66.)

Birch Leaf

(Created 2018)

This is a minor variation of the previous design, Aspen Leaf.

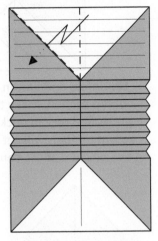

1. Start with Step 12 of Aspen Leaf, but pleat and tuck the top only.

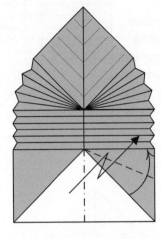

2. Make the angular pleat while bisecting the angle shown.

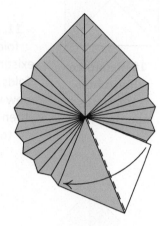

3. Fold white flap to the left.

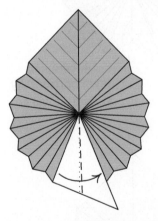

4. Fold white flap just past the vertical, shown dotted.

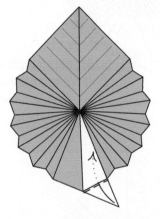

5. Fold the tip and tuck in the opening of the flap to lock.

6. Turn over.

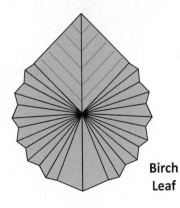

Birch Leaf

6 ◈ Octagonal Collapse Flowers

Top: Abstract Flowers (p. 88). Bottom: Gaillardia and Variations (p. 81).

Gazania

(Created 2013)

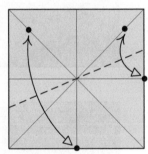

Use duo paper of grade no heavier than *kami*. Kraft works well. Floral colors are desired.

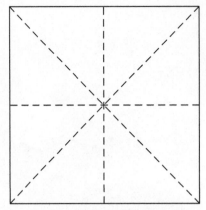

1. Start with 8" or larger square. Fold both diagonals and book folds and unfold.

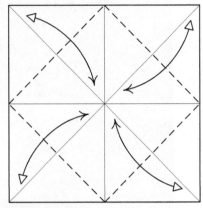

2. Blintz fold and unfold.

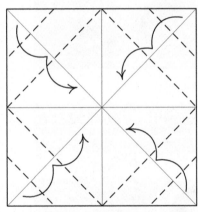

3. Fold each corner to blintz fold, then refold blintz folds.

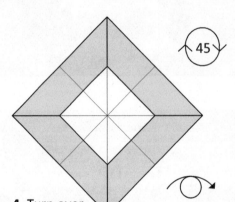

4. Turn over and rotate.

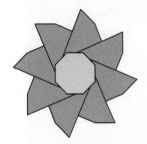

5. Fold to bisect angle, following the dots as guides. Unfold.

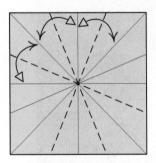

6. Repeat previous step 3 more times.

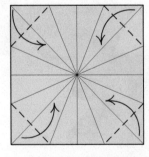

7. Fold corners where shown to get an octagon.

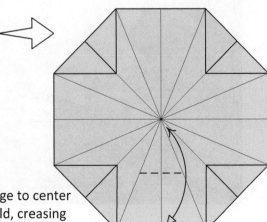

8. Fold edge to center and unfold, creasing only where shown.

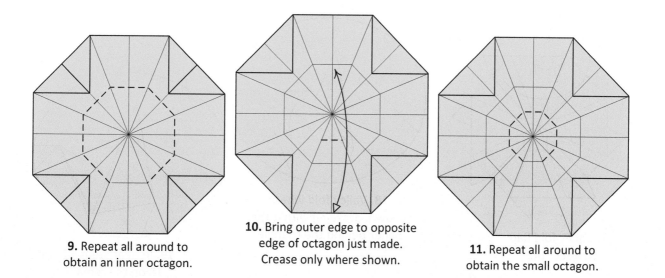

9. Repeat all around to obtain an inner octagon.

10. Bring outer edge to opposite edge of octagon just made. Crease only where shown.

11. Repeat all around to obtain the small octagon.

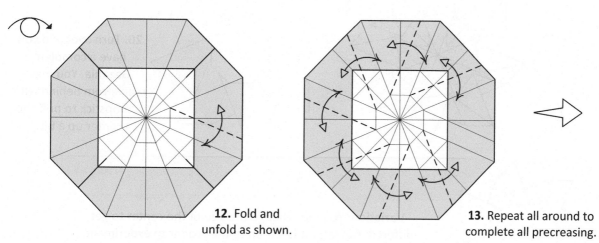

12. Fold and unfold as shown.

13. Repeat all around to complete all precreasing.

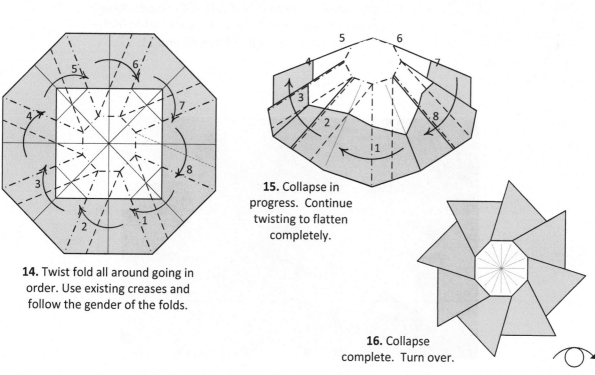

14. Twist fold all around going in order. Use existing creases and follow the gender of the folds.

15. Collapse in progress. Continue twisting to flatten completely.

16. Collapse complete. Turn over.

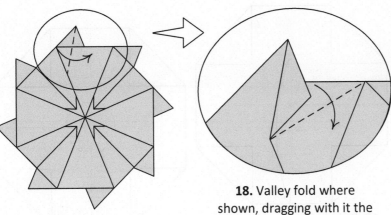

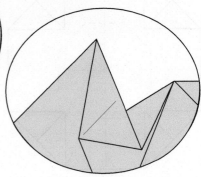

17. Valley fold flap to bisect tip.

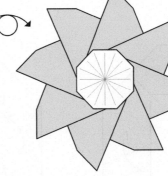

18. Valley fold where shown, dragging with it the paper behind the triangle.

19. Repeat Steps 17 and 18 on the remaining tips to finish all petals.

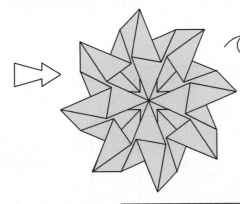

20. Turn over and you have a complete **Gazania**. You may reach from behind with a chopstick to puff the center up a bit.

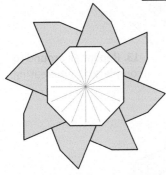

Variation: You may change the size of the center to get different flowers. It is left up to the reader to experiment. The only point to keep in mind is that the outer octagon creased in Steps 8 and 9 should be double that of the inner one creased in Steps 10 and 11.

Gazania and Variation.

Gaillardia

(Created 2013)

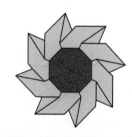

Gaillardia calls for duo paper, yellow or orange shades on one side and dark brown or black on the other. Other floral colors will give the look of different flowers. Paper no heavier than *kami* grade is recommended. Video Instructions can be found in [Ada14].

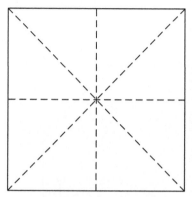

1. Start with 8" or larger square. Fold both diagonals and book folds and unfold.

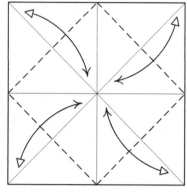

2. Blintz fold and unfold.

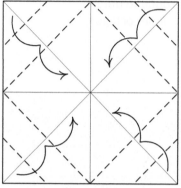

3. Fold each corner to blintz fold and refold blintz fold.

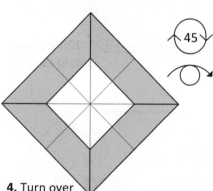

4. Turn over and rotate.

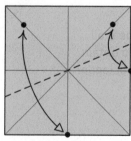

5. Fold to bisect angle, following the dots as guides. Unfold.

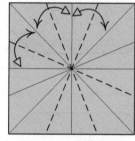

6. Repeat previous step three more times.

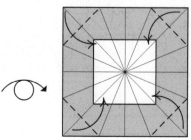

7. Valley fold corners where shown to obtain an octagon.

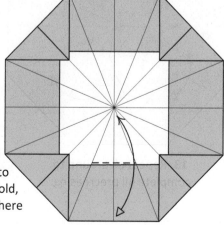

8. Fold edge to center and unfold, creasing only where shown.

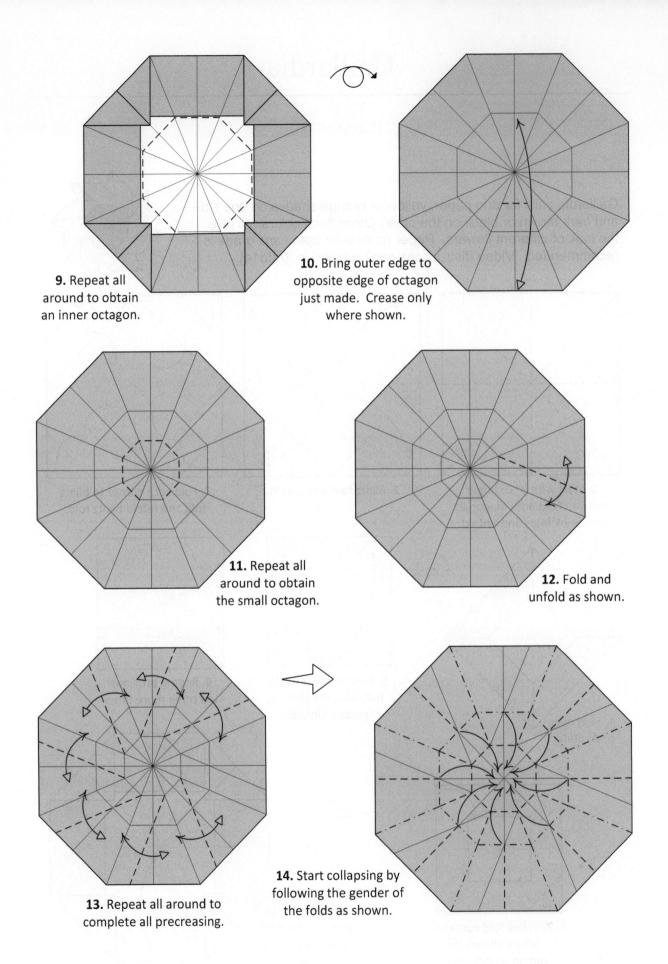

9. Repeat all around to obtain an inner octagon.

10. Bring outer edge to opposite edge of octagon just made. Crease only where shown.

11. Repeat all around to obtain the small octagon.

12. Fold and unfold as shown.

13. Repeat all around to complete all precreasing.

14. Start collapsing by following the gender of the folds as shown.

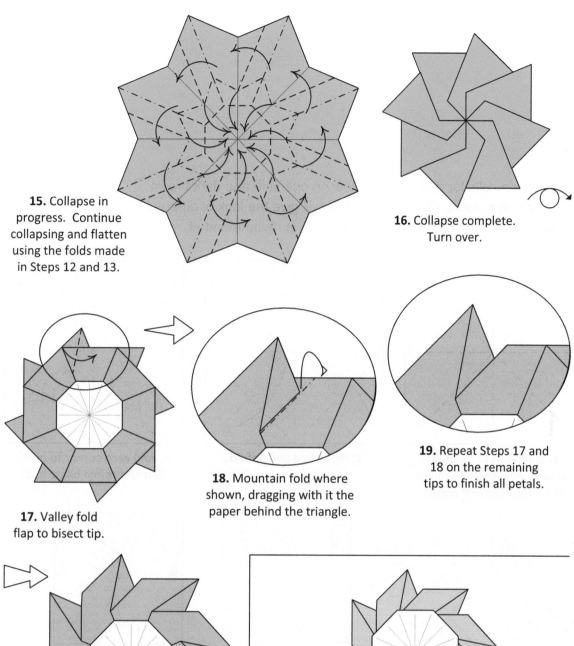

15. Collapse in progress. Continue collapsing and flatten using the folds made in Steps 12 and 13.

16. Collapse complete. Turn over.

17. Valley fold flap to bisect tip.

18. Mountain fold where shown, dragging with it the paper behind the triangle.

19. Repeat Steps 17 and 18 on the remaining tips to finish all petals.

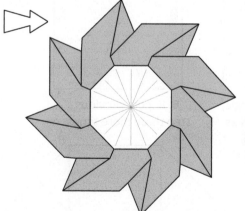

Gaillardia (You may reach from behind with a chopstick to puff the center up a bit.)

Variation: You may change the size of the center to get different looking flowers. It is left up to the reader to experiment. The only point to keep in mind is that the outer octagon creased in Steps 8 and 9 should be double that of the inner one creased in Steps 10 and 11.

(Photo on p. 77.)

Rudbeckia

(Created 2013)

Rudbeckia calls for duo paper, yellow on one side and dark brown or black on reverse. Other floral colors will give the look of different flowers. Paper no heavier than *kami* grade is recommended.

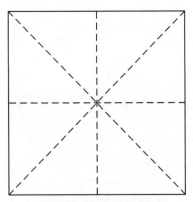

1. Start with 8" or larger square. Fold both diagonals and book folds and unfold.

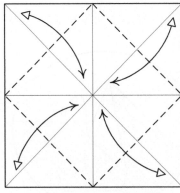

2. Blintz fold and unfold.

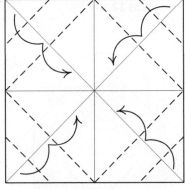

3. Fold each corner to blintz fold and refold blintz fold.

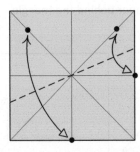

4. Turn over and rotate.

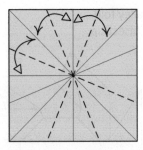

5. Fold to bisect angle, following the dots as guides. Unfold.

6. Repeat previous step three more times.

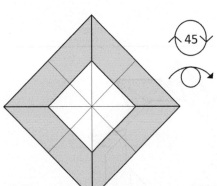

7. Valley fold corners where shown to obtain an octagon.

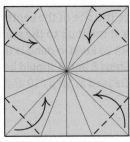

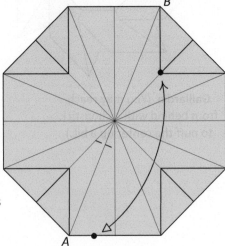

8. Slide vertex *A* up along line *AB* till bottom edge touches flap. Crease only where shown.

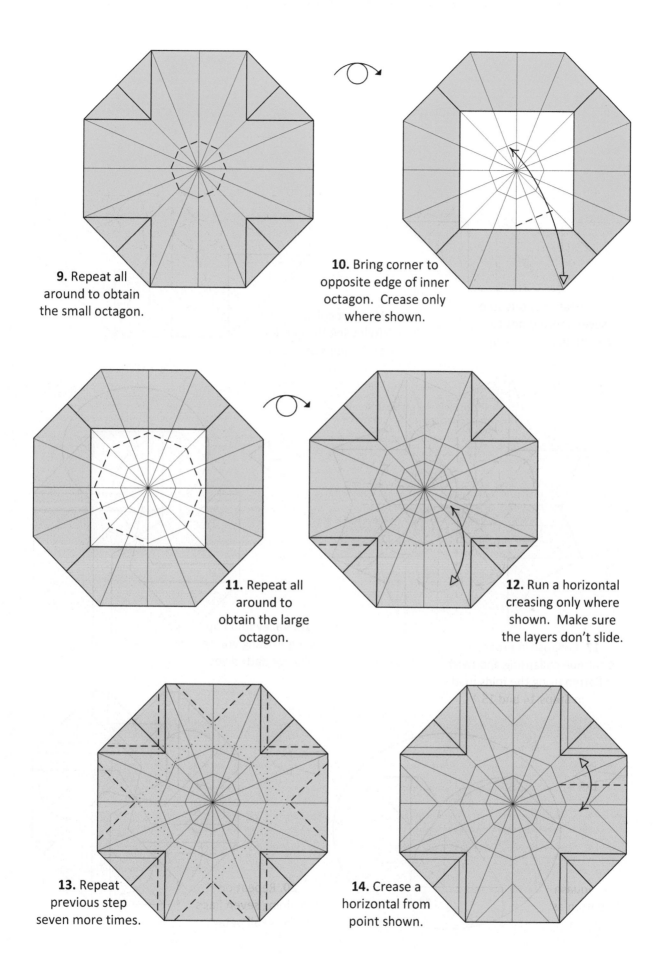

9. Repeat all around to obtain the small octagon.

10. Bring corner to opposite edge of inner octagon. Crease only where shown.

11. Repeat all around to obtain the large octagon.

12. Run a horizontal creasing only where shown. Make sure the layers don't slide.

13. Repeat previous step seven more times.

14. Crease a horizontal from point shown.

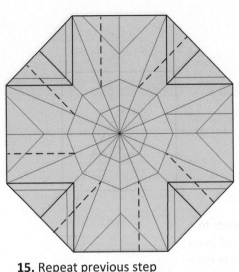

15. Repeat previous step seven more times to complete precreasing.

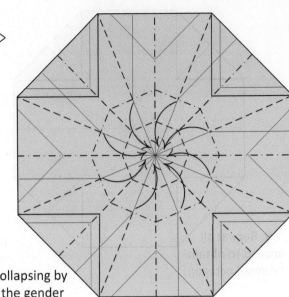

16. Start collapsing by following the gender of the folds as shown.

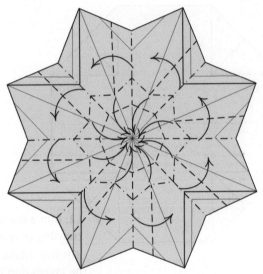

17. Collapse in progress. Continue collapsing and twist to flatten using the folds made in Steps 14 and 15.

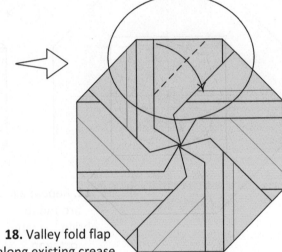

18. Valley fold flap along existing crease. Do not flatten yet.

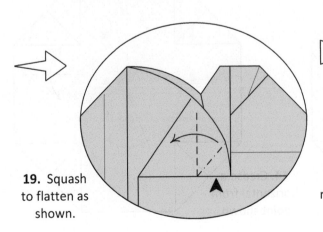

19. Squash to flatten as shown.

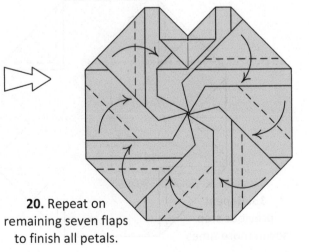

20. Repeat on remaining seven flaps to finish all petals.

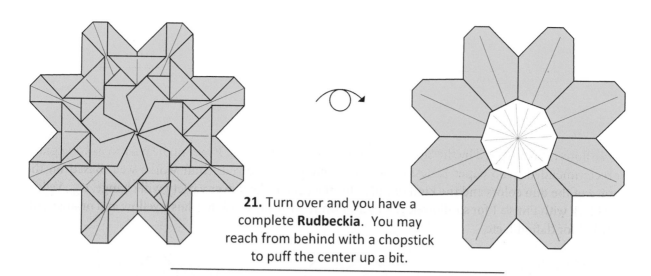

21. Turn over and you have a complete **Rudbeckia**. You may reach from behind with a chopstick to puff the center up a bit.

Variation: You may vary the size of the center to get a different finished look for the flower. It is left up to the reader to experiment. The only point to keep in mind is that the outer octagon creased in Steps 10 and 11 should be double that of the inner one creased in Steps 8 and 9.

Rudbeckia.

Abstract Flowers

(Created 2013)

These flowers are inspired by the Shuzo Fujimoto's Clematis [Bos82]. Paper no heavier than *kami* grade is recommended. Duo paper is desired as color change is involved. Kraft works well. Based on the choice of the duo colors and the kind of finish for the center, the flower variety will change, e.g., yellow and black with Finish 1 for sunflower, yellow and white with any finish for daisy, yellow and orange with Finish 2 for daffodil, etc.

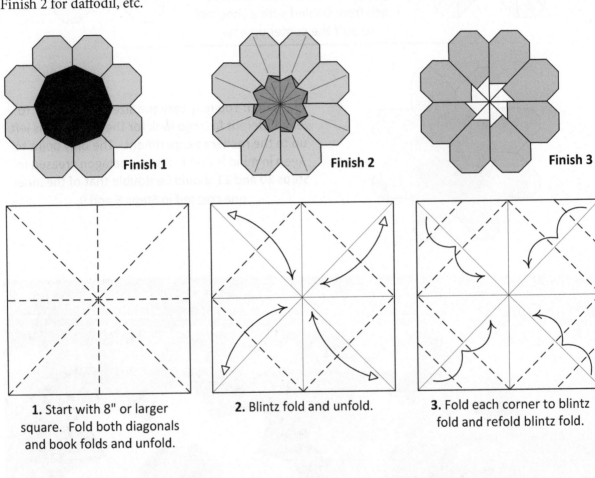

Finish 1 **Finish 2** **Finish 3**

1. Start with 8" or larger square. Fold both diagonals and book folds and unfold.

2. Blintz fold and unfold.

3. Fold each corner to blintz fold and refold blintz fold.

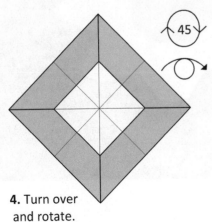

4. Turn over and rotate.

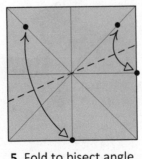

5. Fold to bisect angle, following the dots as guides. Unfold.

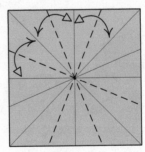

6. Repeat previous step three more times.

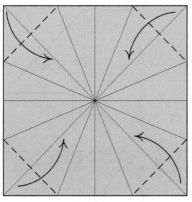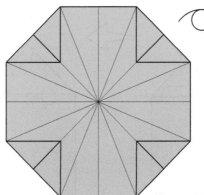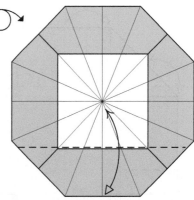

7. Valley fold corners where shown to obtain an octagon.

8. Turn over.

9. Fold bottom edge to center through all layers and unfold. Make sure the layers don't slide.

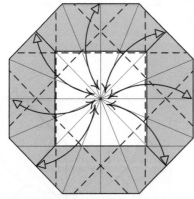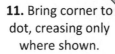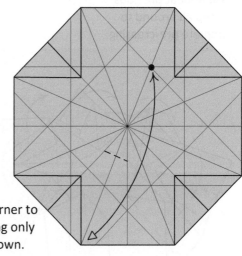

10. Repeat previous step all around the octagon. Turn over.

11. Bring corner to dot, creasing only where shown.

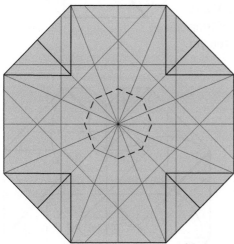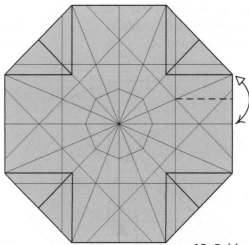

12. Repeat previous step all around to crease a smaller octagon.

13. Fold and unfold, creasing only where shown.

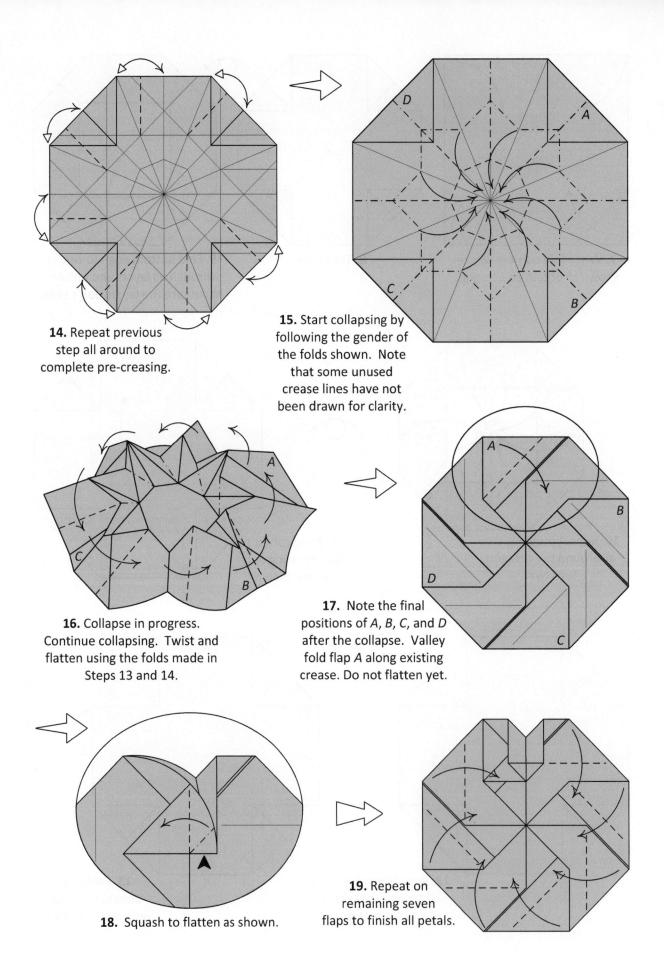

14. Repeat previous step all around to complete pre-creasing.

15. Start collapsing by following the gender of the folds shown. Note that some unused crease lines have not been drawn for clarity.

16. Collapse in progress. Continue collapsing. Twist and flatten using the folds made in Steps 13 and 14.

17. Note the final positions of A, B, C, and D after the collapse. Valley fold flap A along existing crease. Do not flatten yet.

18. Squash to flatten as shown.

19. Repeat on remaining seven flaps to finish all petals.

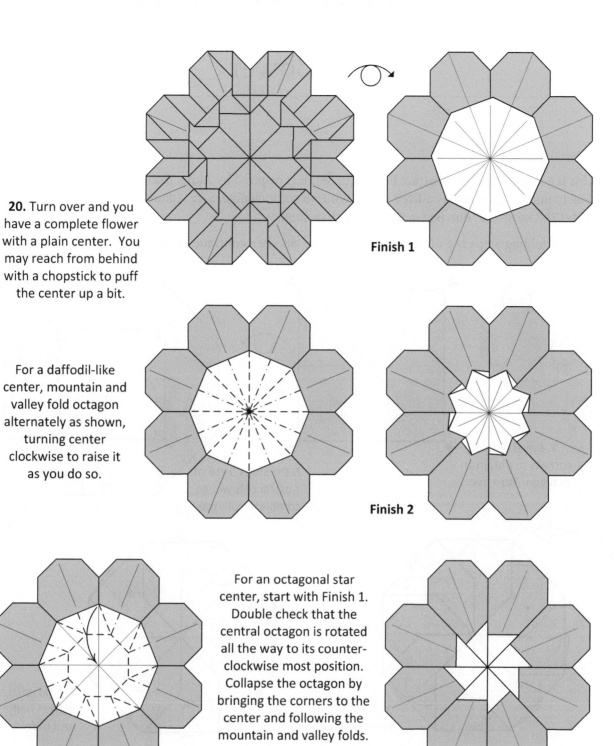

20. Turn over and you have a complete flower with a plain center. You may reach from behind with a chopstick to puff the center up a bit.

Finish 1

For a daffodil-like center, mountain and valley fold octagon alternately as shown, turning center clockwise to raise it as you do so.

Finish 2

For an octagonal star center, start with Finish 1. Double check that the central octagon is rotated all the way to its counter-clockwise most position. Collapse the octagon by bringing the corners to the center and following the mountain and valley folds.

Finish 3

(Photo on p. 77.)

Note: It is possible to design with greater paper efficiency, i.e., to obtain a larger flower starting with the same size paper. However, the designs have been presented the way they are for two main reasons: elegance of diagramming, and a neat finish on the reverse side. If paper efficiency is important to you, you are encouraged to experiment for yourself.

Sunflower

(Created 2015)

This is a variation of my Abstract Flower (p. 88). It is inspired by Akira Yoshizawa's Sunflower [Yos84], which, unlike this one, has a fixed center with no color change. Sara Adams has video instructions in [Ada17]. Use 8″–10″ duo paper, no heavier than *kami*.

Start by folding Steps 1–6 of Abstract Flower on p. 88, then continue as below.

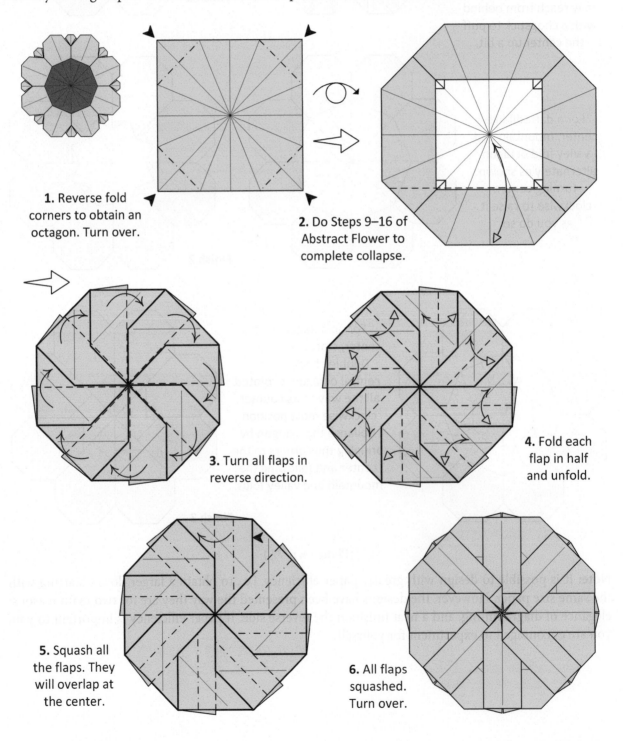

1. Reverse fold corners to obtain an octagon. Turn over.

2. Do Steps 9–16 of Abstract Flower to complete collapse.

3. Turn all flaps in reverse direction.

4. Fold each flap in half and unfold.

5. Squash all the flaps. They will overlap at the center.

6. All flaps squashed. Turn over.

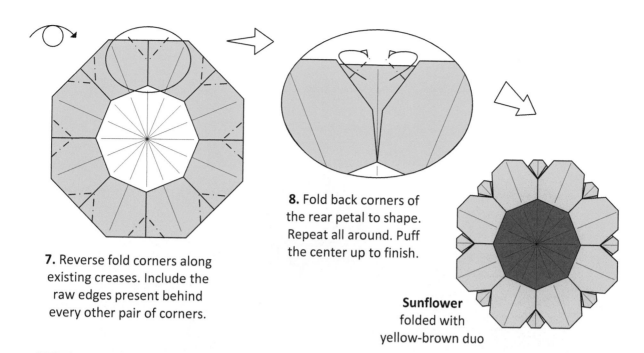

7. Reverse fold corners along existing creases. Include the raw edges present behind every other pair of corners.

8. Fold back corners of the rear petal to shape. Repeat all around. Puff the center up to finish.

Sunflower
folded with
yellow-brown duo

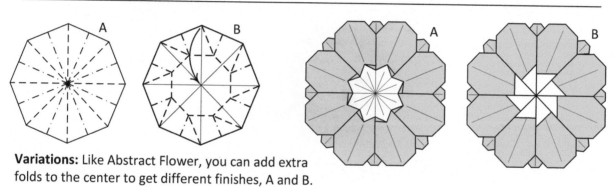

Variations: Like Abstract Flower, you can add extra folds to the center to get different finishes, A and B.

Sunflower folded with yellow and black duo paper.

Octagonal Collapse flowers folded with Corona Harmony Paper. Clockwise from 12 o'clock position: Rudbeckia, Gaillardia, Abstract Flower Finish 2, and Gazania.

If you are using Corona Harmony Paper, you may omit the color changing Steps 2, 3, and 4 for each of the flowers and then continue with the rest of the Steps with the white side of the paper facing you. This reduces manipulating layers of paper, although achieving color change by origami rather than by using special paper is more desirable in pure origami.

You may also try the Sunflower from Corona Harmony paper. In Step 1 of the Sunflower, simply valley fold the corners instead of reverse folding. Also in Step 7 do not worry about raw edges as we do not have any extra layers to manipulate.

Transposing the Flowers to Other Polygons

It would be interesting to make the flowers presented in this chapter from other regular polygons with five or more sides, using the same transposing concept as applied in the diagrams for Traditional and Five Petal Lilies (p. 15).

Left: Gaillardia and Rudbeckia from hexagons. Right: Rudbeckias from nonagon (nine sides) and decagon (ten sides), and Gazania from hendecagon (11 sides).

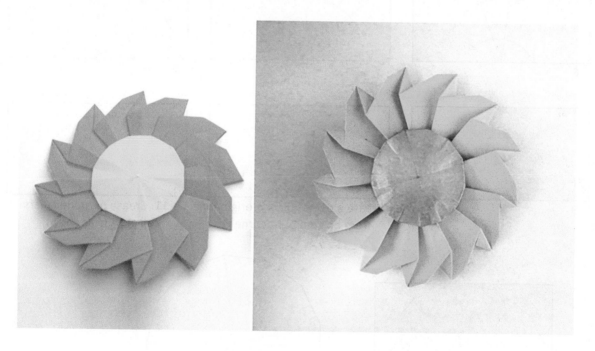

Gaillardias from dodecagon (12 sides) and tridecagon (13 sides).

For fun, we will illustrate how to fold an approximate tridecagon (13-sided polygon) and fold a Gaillardia from it. That should give you an idea of how to fold all the different flowers in this chapter from other regular polygons.

Folding a Tridecagon

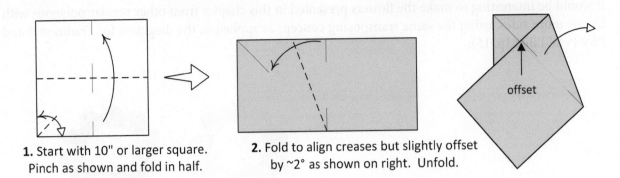

1. Start with 10" or larger square. Pinch as shown and fold in half.

2. Fold to align creases but slightly offset by ~2° as shown on right. Unfold.

offset

In Steps 3–9 below, carefully valley fold and unfold as shown, making sure that all the creases pass through the center. Accuracy is highly crucial.

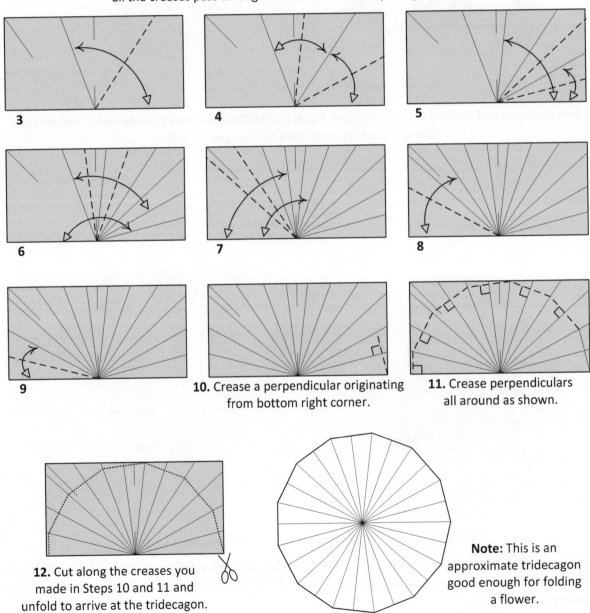

3

4

5

6

7

8

9

10. Crease a perpendicular originating from bottom right corner.

11. Crease perpendiculars all around as shown.

12. Cut along the creases you made in Steps 10 and 11 and unfold to arrive at the tridecagon.

Note: This is an approximate tridecagon good enough for folding a flower.

Explaining the Approach to My Origami Method of Folding a Tridecagon

I wanted to find an easy origami way to make a tridecagon using origami methods. I looked at a few existing methods, such as [Ger08] and [Che14], and thought there must be an easier way. Though my result is approximate, it is good enough for origami folding purposes and the simplicity of my approach may be of interest to some people. The drawback of the method is that it cannot be generalized for all polygons.

After you've made the tridecagon, you may transpose most origami designs based on other regular polygons on to it, for making 13-petaled flowers or other 13-sided geometric objects.

Basic Properties of a Regular Tridecagon

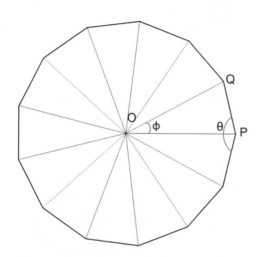

Number of edges = 13

Number of vertices = 13

Internal angle $\theta \approx 152.31°$

For each sector, angle at the center,
$\phi = 360°/13 \approx 27.69°$

I have used a "solve by angle" approach to arrive at the method. I studied if there are any even multiples of ϕ ($\approx 27.69°$, as above) that can be easily obtained by origami.

$\phi \times 2 \approx 55.38°$ — not suitable

$\phi \times 4 \approx 110.76°$ — close enough to 112.5°, which is $90° + 22.5°$, and these angles, in turn, can be achieved with origami. The offset of $\approx 2°$ in Step 2 of my method on the previous page, is to take care of the difference between 112.5° and 110.76°, which is 1.74°.

Once we have made the crease at $\approx 110.76°$ to the X axis, we can simply divide the angle into four parts, which is easy in origami, to get our desired ϕ. This has been done in Steps 3 and 4. After that, it's only a matter of replicating the angle all around the center and finding the edges of the tridecagon such that we get the largest possible one from the starting paper.

Thirteen Petalled Gaillardia

(Created 2014)

Start with a tridecagon as on p. 96. Steps 1–4 are for color change. If you do not want color change, you may start from Step 5, color side up, using a smaller tridecagon.

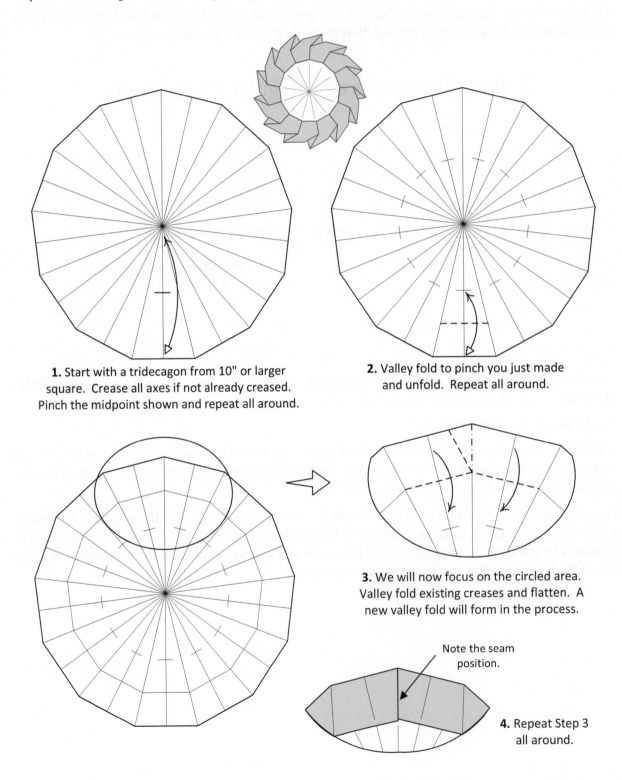

1. Start with a tridecagon from 10" or larger square. Crease all axes if not already creased. Pinch the midpoint shown and repeat all around.

2. Valley fold to pinch you just made and unfold. Repeat all around.

3. We will now focus on the circled area. Valley fold existing creases and flatten. A new valley fold will form in the process.

Note the seam position.

4. Repeat Step 3 all around.

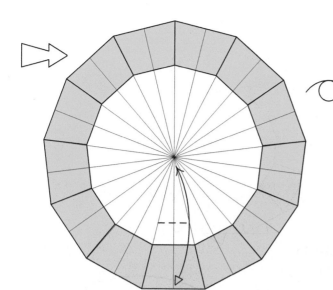

5. Bring edge to center. Valley fold where shown and unfold. Repeat all around to obtain an inner tridecagon.

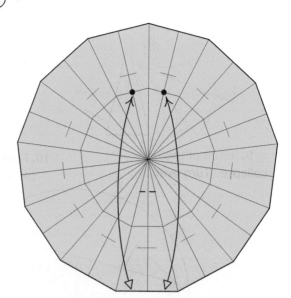

6. Bring bottom edge to dots and make a small crease where shown. Repeat all around to obtain a smaller tridecagon.

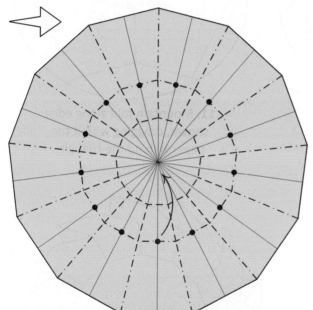

7. Start collapsing following gender of the folds, working from outside to inside. The outer tridecagon is mountain folded and the inner one is valley folded. The axes are part mountain and part valley folded. The dotted points will eventually approach the center.

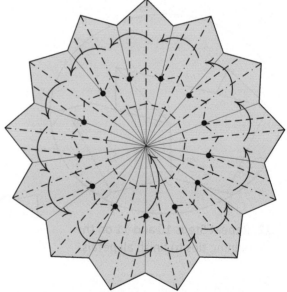

8. Collapse in progress. Continue collapsing and flatten counterclockwise with the new valley folds that form automatically to arrive at the next step. Note that these valley folds are parallel to the axes and may be pre-creased if you wish.

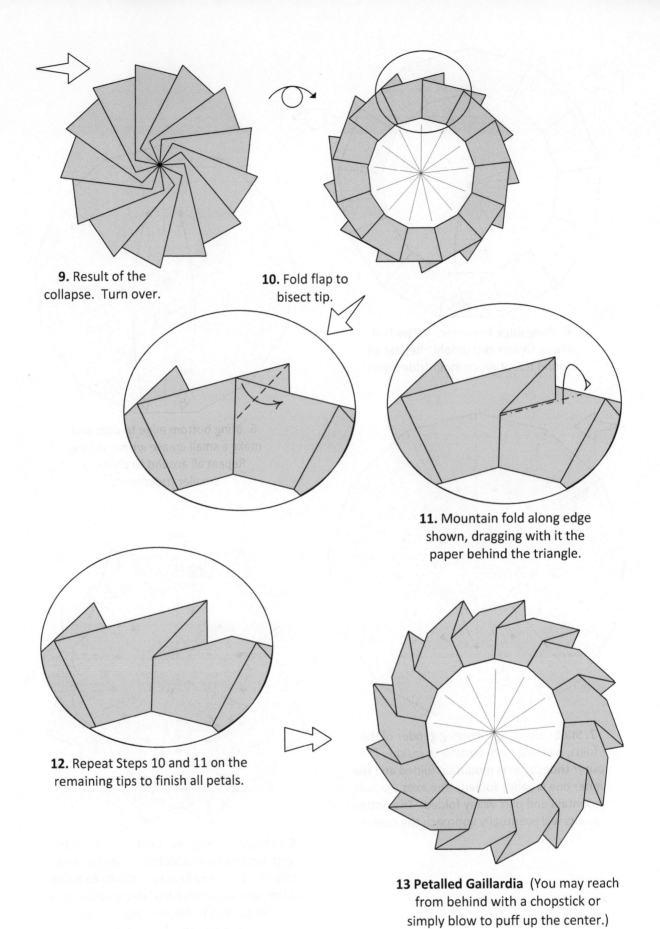

9. Result of the collapse. Turn over.

10. Fold flap to bisect tip.

11. Mountain fold along edge shown, dragging with it the paper behind the triangle.

12. Repeat Steps 10 and 11 on the remaining tips to finish all petals.

13 Petalled Gaillardia (You may reach from behind with a chopstick or simply blow to puff up the center.)

(Photo on p. 95.)

7 ◈ Recursive Designs

Fractal Sakura (next page).

Reverse side of Fractal Sakura, and Octospiral (p. 110).

Fractal Sakura

(Created 2012)

This model is inspired by Shuzo Fujimoto's Hydrangea [Fuj10] and Roman Diaz's Fractal Flower [Dia12]. Start with a Four-Sink Base (p. 30) from 12″ or larger paper and then continue as below.

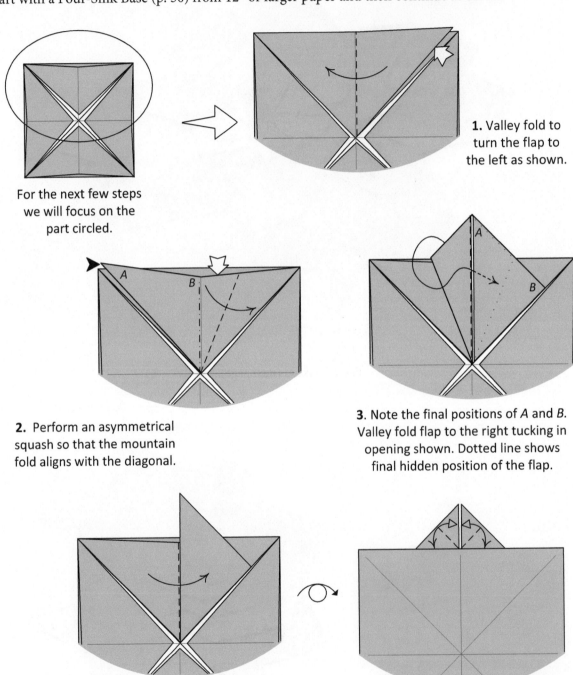

For the next few steps we will focus on the part circled.

1. Valley fold to turn the flap to the left as shown.

2. Perform an asymmetrical squash so that the mountain fold aligns with the diagonal.

3. Note the final positions of A and B. Valley fold flap to the right tucking in opening shown. Dotted line shows final hidden position of the flap.

4. Turn left flap to the right and repeat Steps 2 and 3, mirrored.

5. Fold and unfold the two tips.

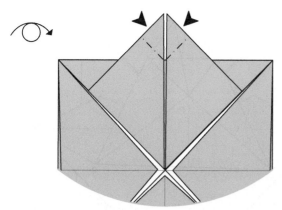

6. Reverse fold along creases you just made.

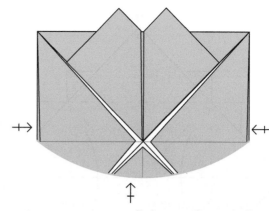

7. Repeat Steps 1–6 on the remaining three sides.

8. Turn over.

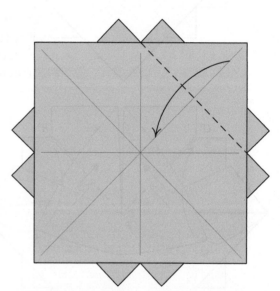

9. Valley fold corner while petal folding the layers getting pulled from behind. See next figure for the resulting look.

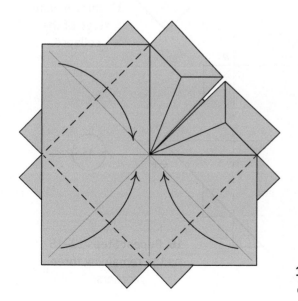

10. Repeat Step 9 on remaining three corners.

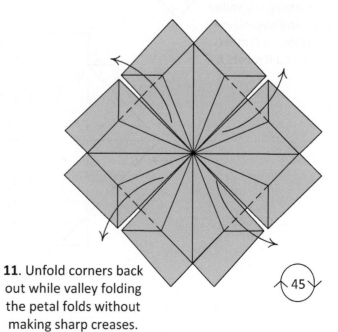

11. Unfold corners back out while valley folding the petal folds without making sharp creases.

45

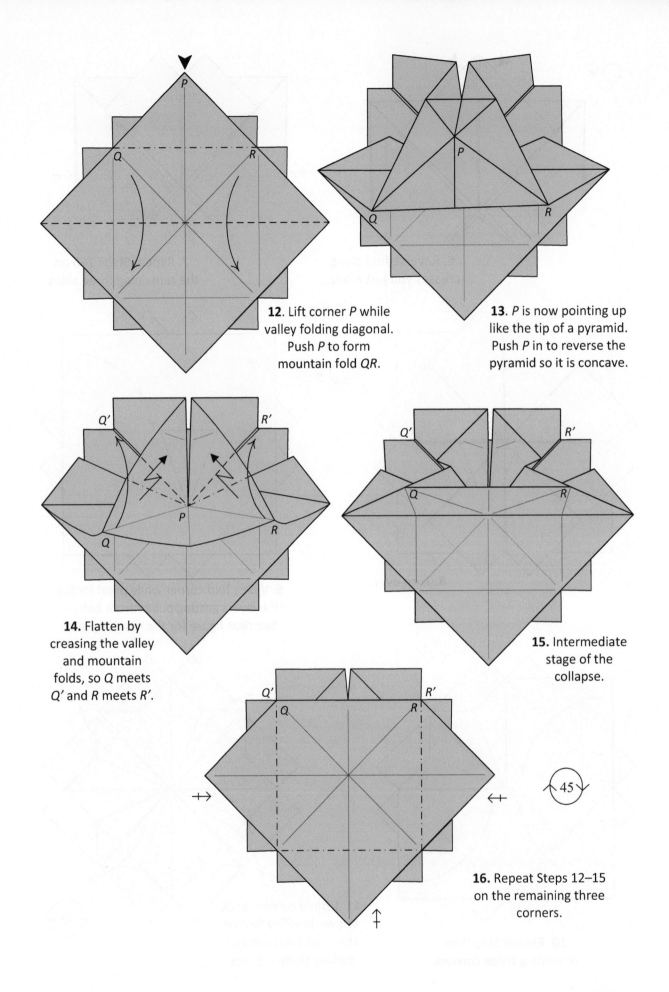

12. Lift corner *P* while valley folding diagonal. Push *P* to form mountain fold *QR*.

13. *P* is now pointing up like the tip of a pyramid. Push *P* in to reverse the pyramid so it is concave.

14. Flatten by creasing the valley and mountain folds, so *Q* meets *Q′* and *R* meets *R′*.

15. Intermediate stage of the collapse.

16. Repeat Steps 12–15 on the remaining three corners.

Fractal Sakura

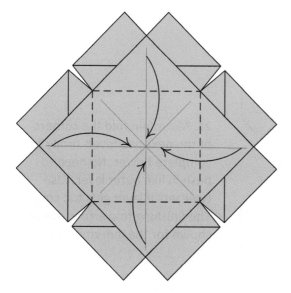

17. Repeat Steps 9 and 10 on corners shown.

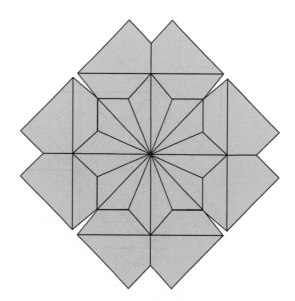

18. This is an unfinished two-level flower. Repeat Steps 11–17 to add another level.

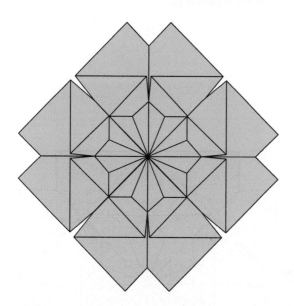

19. This is an unfinished three-level flower. Continue adding levels for as long as you like, or until you and/or the paper reach physical limitations.

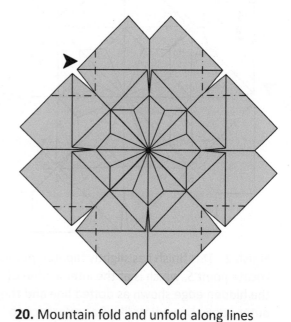

20. Mountain fold and unfold along lines shown and then "close sink" all eight corners. This step is not pleasant, but any crushed paper gets hidden within. Using a tool such as a pointy chopstick helps. Note that the close sinks are optional and you may leave the folds as mountain folds, in which case the reverse side will not be as neat.

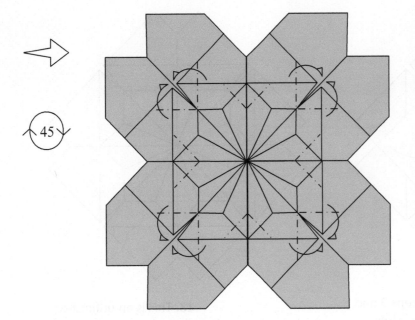

21. Mountain fold the corners of the next two levels firmly to complete flower. No need to push in like outer layer. Use tweezers if layer gets too small. The finishing fold arrows are not shown for the innermost level.

Two other possible finishes with tapered petals.

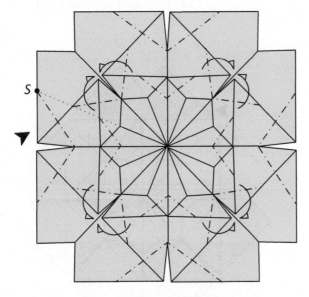

Finish 2: This finish has slightly tapered petals. Locate point *S*, which is at the intersection of the hidden edge shown as dotted line and the outer edge. Use points like *S* as reference to make all eight mountain folds. As explained in Steps 20 and 21, push in the corners for the outer layer, but simply mountain fold crisply for the inner layers. The finishing fold arrows are not shown for the innermost level.

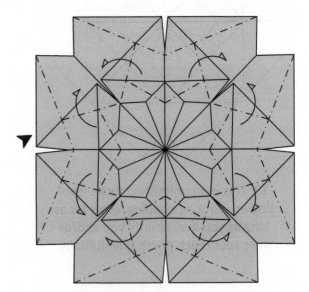

Finish 3: This finish has pointy petals. As explained in Steps 20 and 21, push in the corners for the outer layer, but simply mountain fold crisply for the inner layers. The finishing fold arrows are not shown for the innermost level.

    Fractal Sakura

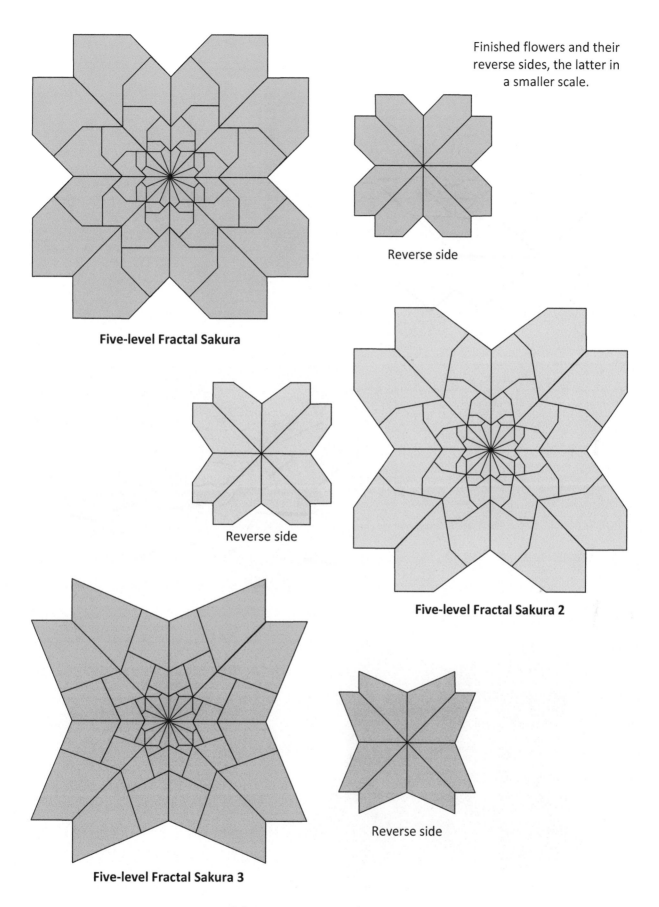

Finished flowers and their reverse sides, the latter in a smaller scale.

Five-level Fractal Sakura

Reverse side

Reverse side

Five-level Fractal Sakura 2

Five-level Fractal Sakura 3

Reverse side

(Photos on p. 101 and next page.)

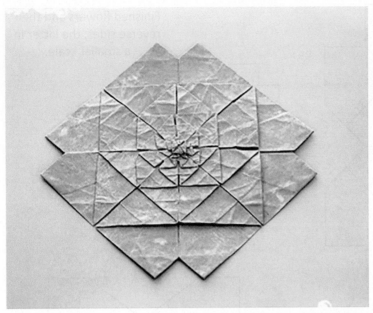

Fractal Sakuras: Top: Unfinished; Middle: Finish 2; Bottom: Finish 3. Finish 1 photo on p. 101.

Fractal Sakura

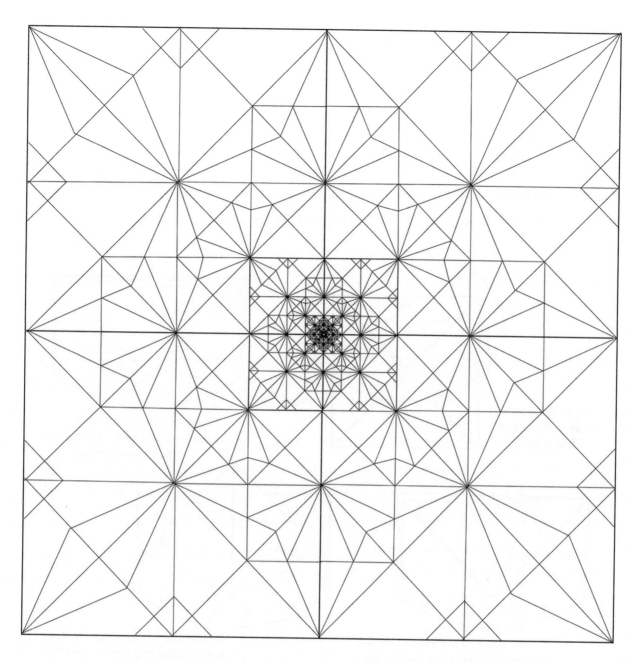

Crease Pattern for Fractal Sakura, not showing orientation of the folds.

Crease Pattern, as the name implies, is a pattern that you see after unfolding a folded piece of origami. Sometimes the crease pattern is so beautiful that it may be a piece of art in its own right. For a fractal design, the crease pattern in turn is a fractal, not surprisingly. I am choosing my Fractal Sakura crease pattern as an example to illustrate, though it is not meant for a folding guide.

Many artists prefer to release crease patterns as opposed to how-to diagrams. Folding from these is obviously much more challenging and they are only meant for the experienced. Some interesting read about crease patterns may be found in [Hud11], [Lan15, Lan18], and [ORC19].

Octospiral

(Created 2012)

Octospiral is a simplified version of the Fractal Flower [Dia12] by Roman Diaz. The variation is by Endre Somos [Som07]. Both Roman and Endre have given me their kind permissions to publish the diagrams. I reverse engineered the design from a photo and serialized the steps.

Start with a Four-Sink Base (p. 30) from 12″ or larger paper and continue as below.

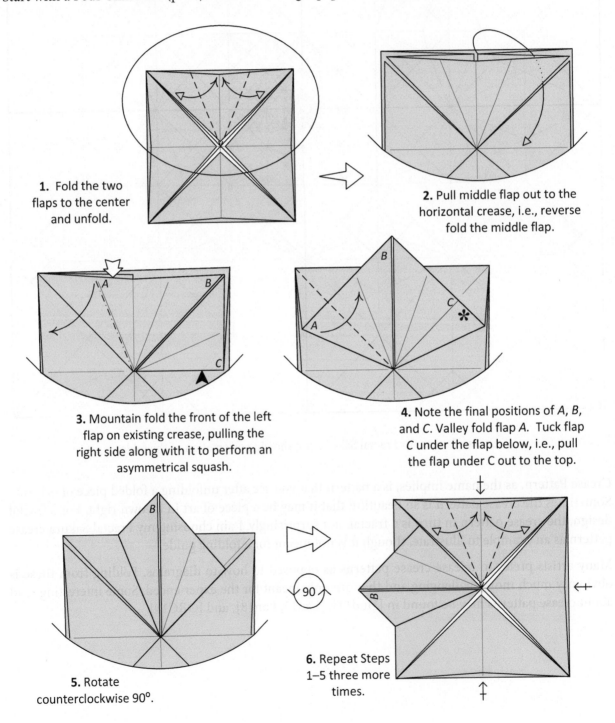

1. Fold the two flaps to the center and unfold.

2. Pull middle flap out to the horizontal crease, i.e., reverse fold the middle flap.

3. Mountain fold the front of the left flap on existing crease, pulling the right side along with it to perform an asymmetrical squash.

4. Note the final positions of *A*, *B*, and *C*. Valley fold flap *A*. Tuck flap *C* under the flap below, i.e., pull the flap under *C* out to the top.

5. Rotate counterclockwise 90°.

6. Repeat Steps 1–5 three more times.

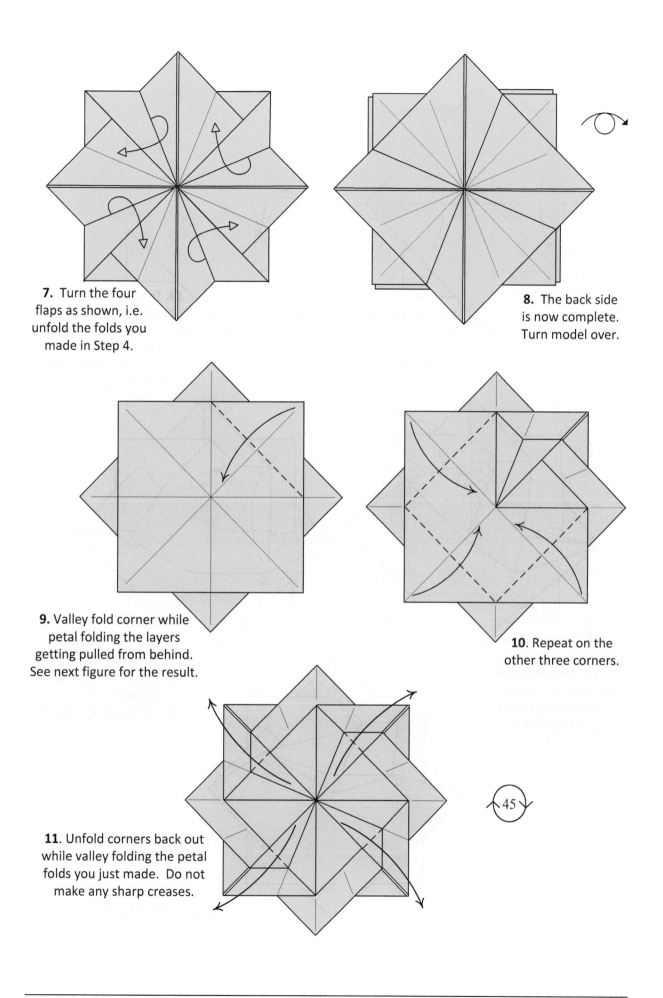

7. Turn the four flaps as shown, i.e. unfold the folds you made in Step 4.

8. The back side is now complete. Turn model over.

9. Valley fold corner while petal folding the layers getting pulled from behind. See next figure for the result.

10. Repeat on the other three corners.

45

11. Unfold corners back out while valley folding the petal folds you just made. Do not make any sharp creases.

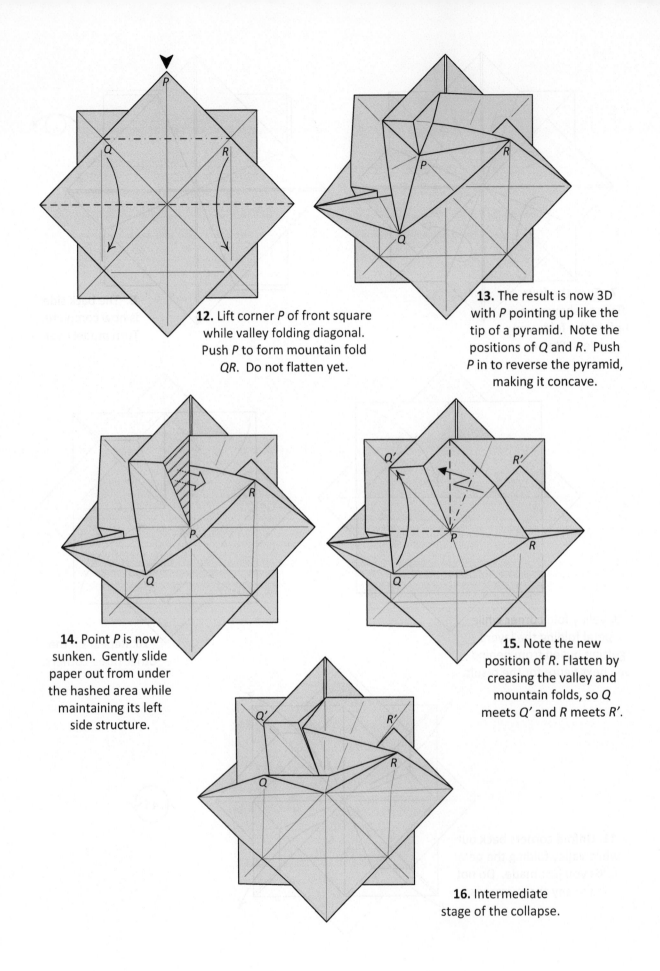

12. Lift corner *P* of front square while valley folding diagonal. Push *P* to form mountain fold *QR*. Do not flatten yet.

13. The result is now 3D with *P* pointing up like the tip of a pyramid. Note the positions of *Q* and *R*. Push *P* in to reverse the pyramid, making it concave.

14. Point *P* is now sunken. Gently slide paper out from under the hashed area while maintaining its left side structure.

15. Note the new position of *R*. Flatten by creasing the valley and mountain folds, so *Q* meets *Q'* and *R* meets *R'*.

16. Intermediate stage of the collapse.

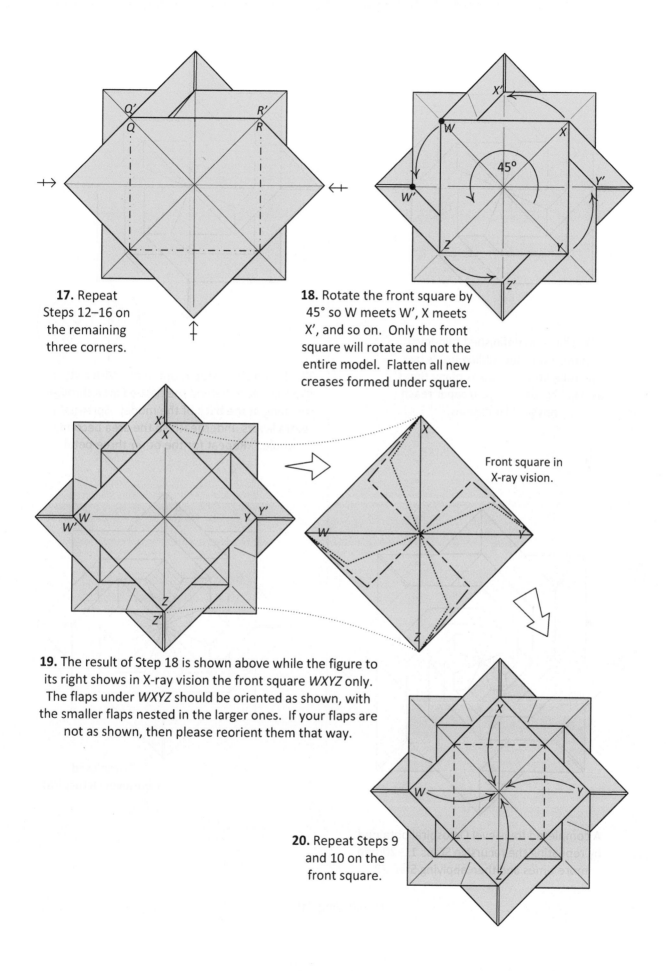

17. Repeat Steps 12–16 on the remaining three corners.

18. Rotate the front square by 45° so W meets W', X meets X', and so on. Only the front square will rotate and not the entire model. Flatten all new creases formed under square.

Front square in X-ray vision.

19. The result of Step 18 is shown above while the figure to its right shows in X-ray vision the front square *WXYZ* only. The flaps under *WXYZ* should be oriented as shown, with the smaller flaps nested in the larger ones. If your flaps are not as shown, then please reorient them that way.

20. Repeat Steps 9 and 10 on the front square.

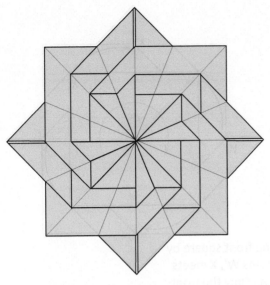

21. This is an unfinished two-level spiral. Continue adding levels by repeating Steps 11–20 as many times as you like, or until you/paper reach physical limitations.

22. This finishing step is optional. With a stylus-like tool reach behind the hashed area through opening at the back of the model. Spread the extra layers underneath so the area becomes "whole." Repeat for the other three petals.

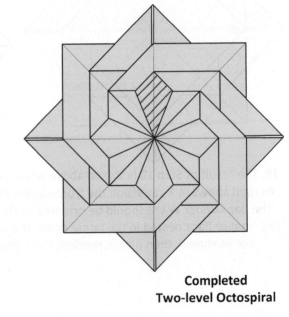

Completed Two-level Octospiral

A completed four-level Octospiral obtained by repeating the recursive Steps 11–20 two more times and then applying Step 22.

(Photo on p. 101.)

Two- and three-level Whirls (next page).

Hydrangea with Leaves (p. 119).

Whirl

(Created 2017)

Whirl is a recursive design inspired by Chris Palmer's [Pam08] Flower Tower [Sha02], experience with which is highly recommended. The same de-creeping technique of excavating paper to add new levels is used. Theoretically, you can add an infinite number of layers, but you will have to stop at some point due to the physical limitations of the paper or the human fingers. When folded from a 9″ square, the minimum size recommended, three levels seem to be sufficient. Folding any more does not enhance the beauty.

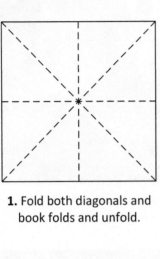

1. Fold both diagonals and book folds and unfold.

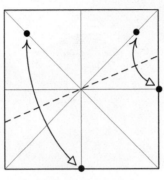

2. Fold to bisect angle, following the dots as guides. Unfold.

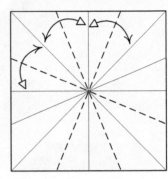

3. Repeat previous step three more times, finding additional guides as you fold.

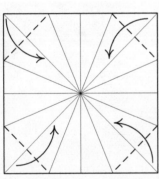

4. Fold corners to obtain an octagon.

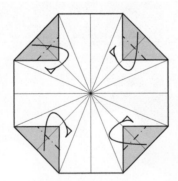

5. Fold the colored triangles in or trim them off entirely.

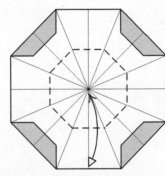

6. Fold edge to center. Repeat all around to get an inner octagon.

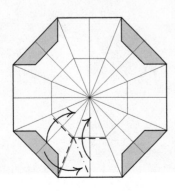

7. Fold using existing creases and the new mountain fold. Do not flatten completely.

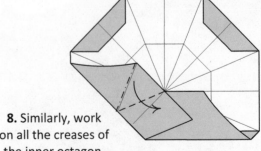

8. Similarly, work on all the creases of the inner octagon, going clockwise.

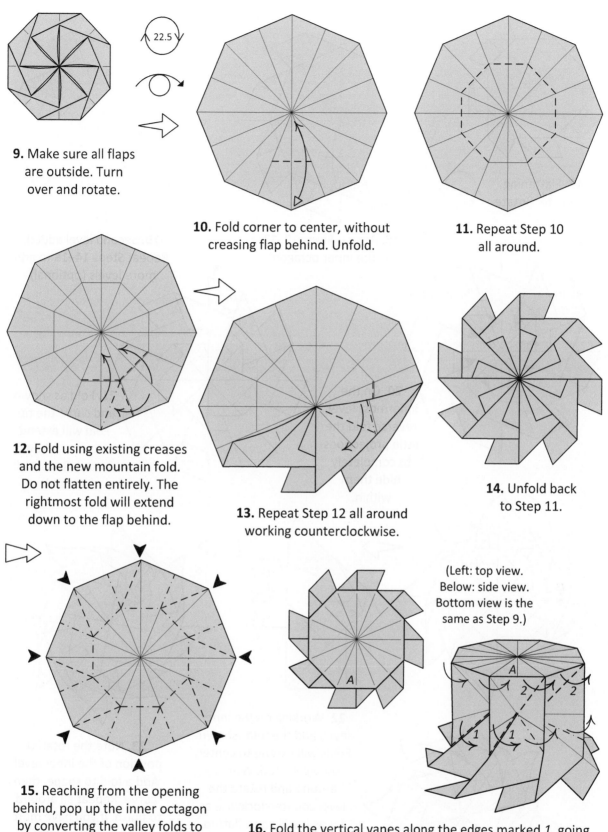

9. Make sure all flaps are outside. Turn over and rotate.

10. Fold corner to center, without creasing flap behind. Unfold.

11. Repeat Step 10 all around.

12. Fold using existing creases and the new mountain fold. Do not flatten entirely. The rightmost fold will extend down to the flap behind.

13. Repeat Step 12 all around working counterclockwise.

14. Unfold back to Step 11.

(Left: top view. Below: side view. Bottom view is the same as Step 9.)

15. Reaching from the opening behind, pop up the inner octagon by converting the valley folds to mountain. At the same time, push in all corners by reversing some of the existing creases. There will be no new creases in the process.

16. Fold the vertical vanes along the edges marked *1*, going all around. Rotate the top octagon counterclockwise, adding folds marked *2* all around, till it reaches the bottom and doesn't rotate any further. Final rotation will be 45° as indicated by the position of *A* in the next figure.

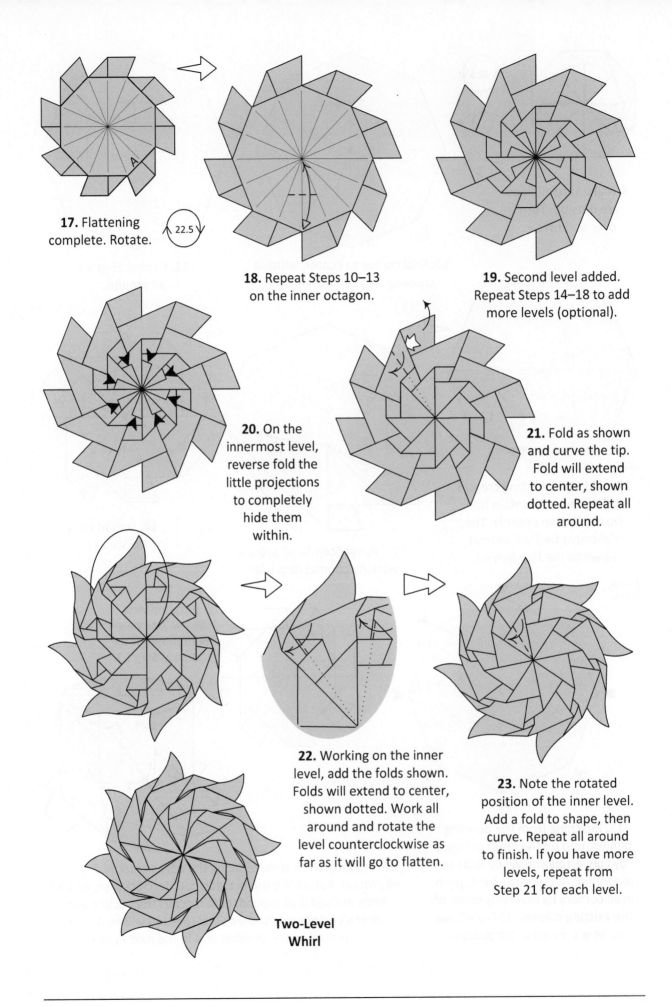

17. Flattening complete. Rotate.

22.5

18. Repeat Steps 10–13 on the inner octagon.

19. Second level added. Repeat Steps 14–18 to add more levels (optional).

20. On the innermost level, reverse fold the little projections to completely hide them within.

21. Fold as shown and curve the tip. Fold will extend to center, shown dotted. Repeat all around.

22. Working on the inner level, add the folds shown. Folds will extend to center, shown dotted. Work all around and rotate the level counterclockwise as far as it will go to flatten.

Two-Level Whirl

23. Note the rotated position of the inner level. Add a fold to shape, then curve. Repeat all around to finish. If you have more levels, repeat from Step 21 for each level.

Hydrangea with Leaves

(Created 2015)

This is an adaptation of Shuzo Fujimoto's Hydrangea [Fuj06, Fuj10]. You get a sturdier and color changed first layer to depict leaves. To fold this design, you must have prior experience with the Hydrangea. Use 8″ or larger paper no heavier than *kami* weight. You can also find video instructions here [Ada15].

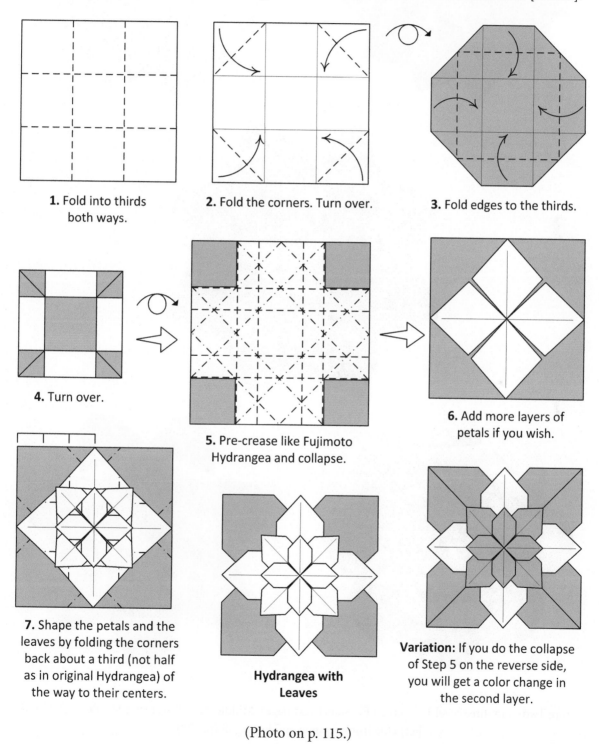

1. Fold into thirds both ways.

2. Fold the corners. Turn over.

3. Fold edges to the thirds.

4. Turn over.

5. Pre-crease like Fujimoto Hydrangea and collapse.

6. Add more layers of petals if you wish.

7. Shape the petals and the leaves by folding the corners back about a third (not half as in original Hydrangea) of the way to their centers.

Hydrangea with Leaves

Variation: If you do the collapse of Step 5 on the reverse side, you will get a color change in the second layer.

(Photo on p. 115.)

Top: Two- and three-level 12-Pointed EZ Stars (next page). Middle: Star Tower (p. 124). Bottom: Floral Perpetua from octagon and heptagon (p. 128).

Hydrangea with Leaves

12-Pointed EZ Star

(Created 2011 by Evan Zodl)

Designed and Diagrammed by Evan Zodl. This model is folded from one uncut square. Start with the white side up.

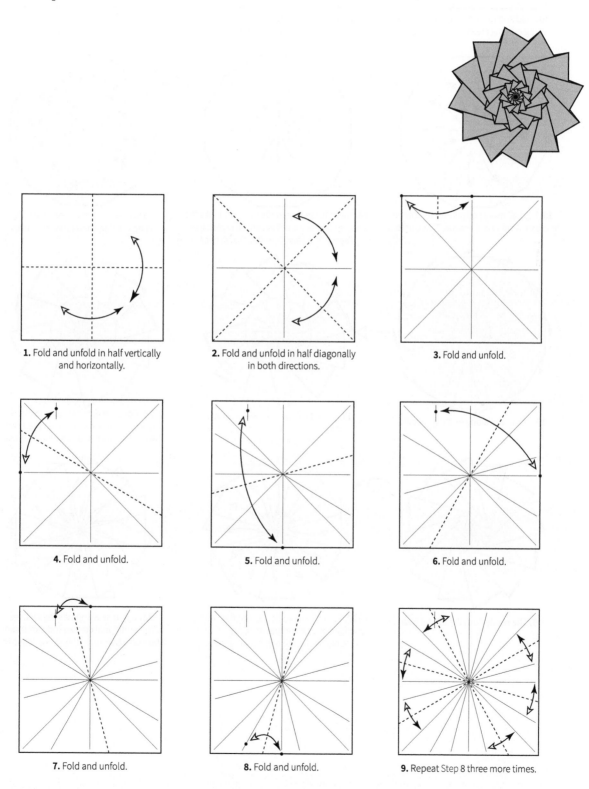

1. Fold and unfold in half vertically and horizontally.

2. Fold and unfold in half diagonally in both directions.

3. Fold and unfold.

4. Fold and unfold.

5. Fold and unfold.

6. Fold and unfold.

7. Fold and unfold.

8. Fold and unfold.

9. Repeat Step 8 three more times.

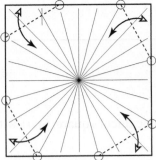

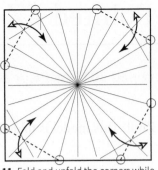

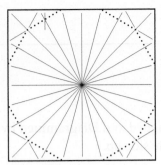

10. Fold and unfold the corners while using existing creases as references.

11. Fold and unfold the corners while using existing creases as references.

12. Cut off the corners along existing creases to create a dodecagon.

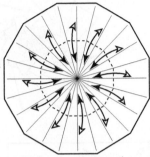

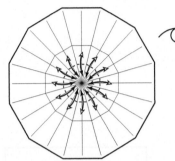

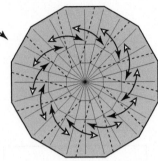

13. Fold all twelve edges into the center to create a small dodecagon. Then, unfold.

14. Fold the creases made in Step 13 into the center to create a smaller central dodecagon. Then, unfold and turn the model over.

15. Create 12 new creases while using existing creases as references.

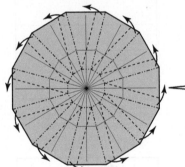

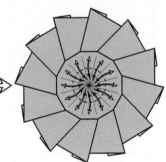

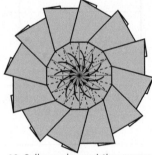

16. Twist fold along existing creases.

17. Fold the edges of the dodecagon into the center. Crease only on the top layer, then unfold.

18. Collapse along existing creases. The additional creases indicated can be made before or during the collapse.

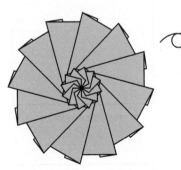

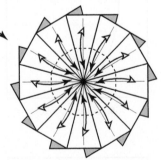

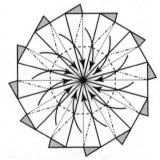

19. The result. Turn the model over.

20. Fold all twelve edges into the center to create a small central dodecagon. Crease only on the top layer, then unfold.

21. Collapse along existing creases. The additional creases indicated can be made before or during the collapse.

12-Pointed EZ Star

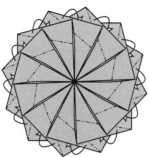

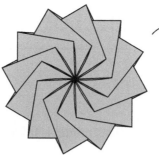

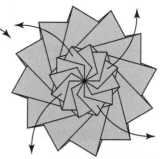

22. Mountain fold the edge of each flap behind, and align it with the edge underneath.

23. The result. Turn the model over.

24. This is a completed two-level EZ Star. To create additional iterations of stars, unfold the central star.

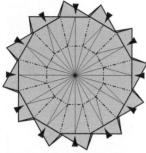

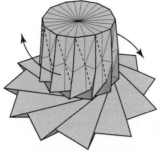

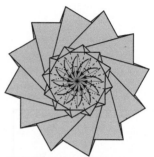

25. Raise the small central dodecagon by creating the indicated mountain folds. This can easily be done by partially unfolding the model. The model will not lie flat.

26. Twist the paper along the indicated creases until the model lies flat.

27. Collapse along existing creases. The additional creases indicated can be made before or during the collapse.

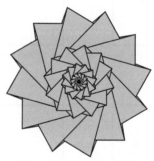

28. This is a completed three-level EZ Star. To create additional iterations of stars, repeat Steps 24–27 on the top star until satisfied.

29. Completed 12-Pointed EZ Star.

Variations: You may fold EZ Star from other polygons as well. Shown are pentagonal, hexagonal, octagonal, and nonagonal versions left up to the reader to try. Original design photos on p. 120.

Star Tower

(Created 2012 by Evan Zodl)

Designed and Diagrammed by Evan Zodl. This model is folded from one uncut square. Start with the white side up.

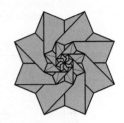

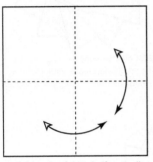

1. Fold and unfold in half vertically and horizontally.

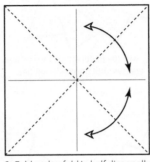

2. Fold and unfold in half diagonally in both directions.

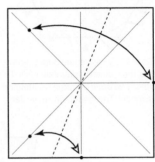

3. Fold and unfold.

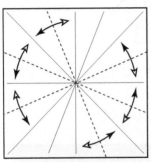

4. Repeat Step 3 three more times.

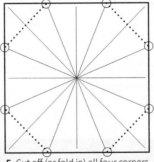

5. Cut off (or fold in) all four corners to create an octagon. Then, turn the model over.

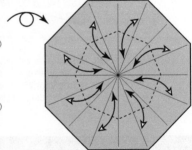

6. Fold all eight corners to the center to create a small central octagon. Then, unfold.

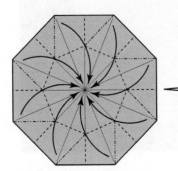

7. Collapse along existing creases. The additional creases indicated can be made before or during the collapse.

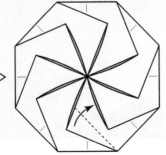

8. Fold up the bottom flap and align it with the raw edges of the next flap.

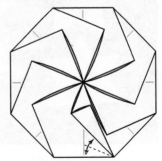

9. Fold and unfold.

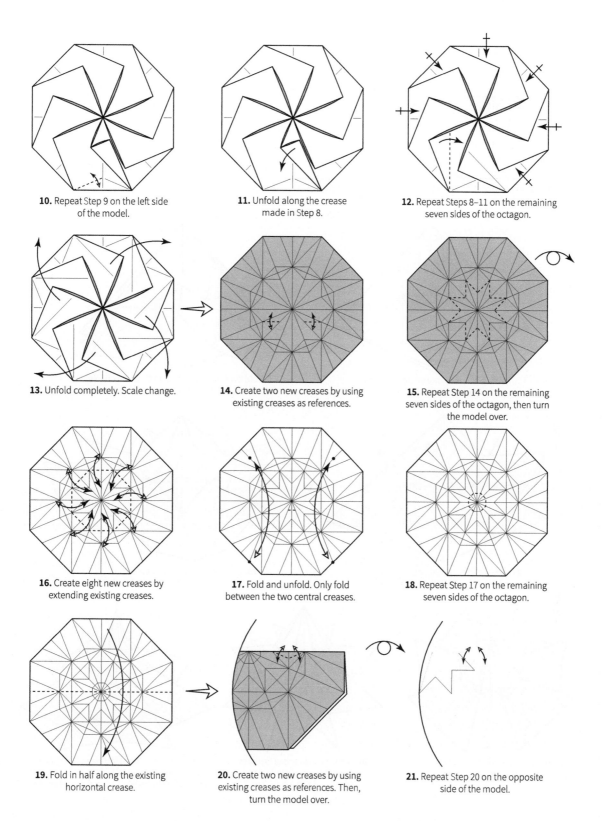

10. Repeat Step 9 on the left side of the model.

11. Unfold along the crease made in Step 8.

12. Repeat Steps 8–11 on the remaining seven sides of the octagon.

13. Unfold completely. Scale change.

14. Create two new creases by using existing creases as references.

15. Repeat Step 14 on the remaining seven sides of the octagon, then turn the model over.

16. Create eight new creases by extending existing creases.

17. Fold and unfold. Only fold between the two central creases.

18. Repeat Step 17 on the remaining seven sides of the octagon.

19. Fold in half along the existing horizontal crease.

20. Create two new creases by using existing creases as references. Then, turn the model over.

21. Repeat Step 20 on the opposite side of the model.

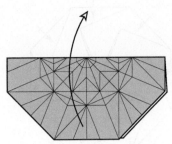

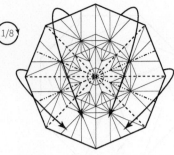

22. The result. Unfold completely.

23. Repeat Steps 19–22 three more times to create the indicated creases. Then, rotate the model.

1/8

24. Collapse flat along existing creases. Steps 25–28 show an enlarged view of the model.

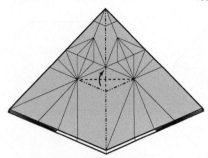

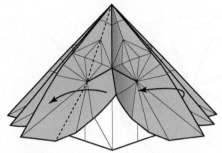

25. Collapse along existing creases. The model will not lie flat.

26. Fold the top central layer to the left along an existing crease while flattening the top right layer.

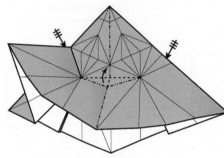

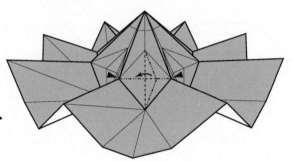

27. Repeat Steps 25–26 on the remaining seven sides of the model.

28. Collapse along existing creases.

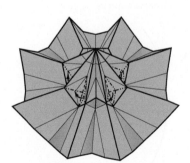

29. Top view. Repeat Step 28 on the remaining seven sides of the model.

30. Twist the top central layers by collapsing along existing creases. The model will not lie flat.

31. Partially flatten the model by twisting along existing creases. Lightly push down on the top of the model to flatten the bottom layers.

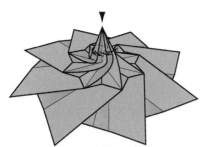

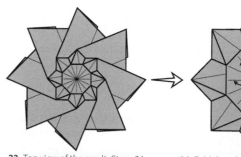

32. Flatten the central point by pushing down on the top of the model and collapsing along existing creases.

33. Top view of the result. Steps 34 and 35 are detailed views of star in the center of the model.

34. Fold the edges of the octagon into the center to create a smaller cental octagon. Crease only on the top layer, then unfold.

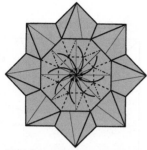

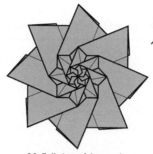

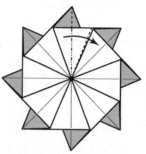

35. Collapse along existing creases. The additional creases indicated can be made before or during the collapse.

36. Full view of the result. Turn the model over.

37. Fold the top layer to the right along existing creases.

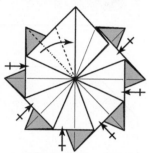

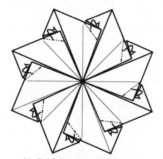

38. Repeat Step 37 on the remaining seven sides.

39. Fold the corners behind to partially lock the model.

40. The result. Turn the model over.

41. Completed Star Tower.

(Photo on p. 120.)

Floral Perpetua

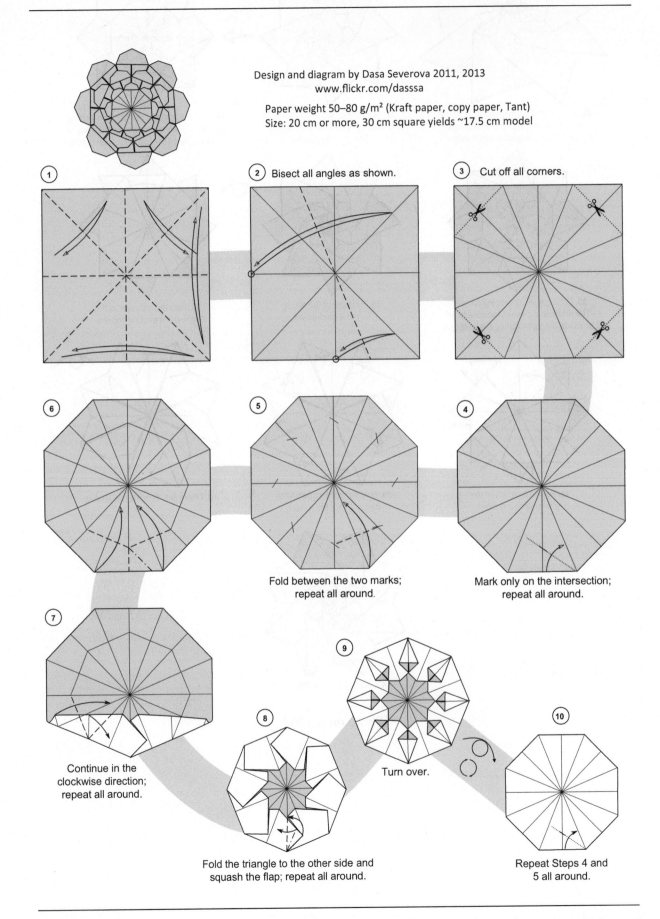

Design and diagram by Dasa Severova 2011, 2013
www.flickr.com/dasssa

Paper weight 50–80 g/m² (Kraft paper, copy paper, Tant)
Size: 20 cm or more, 30 cm square yields ~17.5 cm model

1

2 Bisect all angles as shown.

3 Cut off all corners.

6

5

Fold between the two marks;
repeat all around.

4

Mark only on the intersection;
repeat all around.

7

Continue in the
clockwise direction;
repeat all around.

9

8

Turn over.

10

Fold the triangle to the other side and
squash the flap; repeat all around.

Repeat Steps 4 and
5 all around.

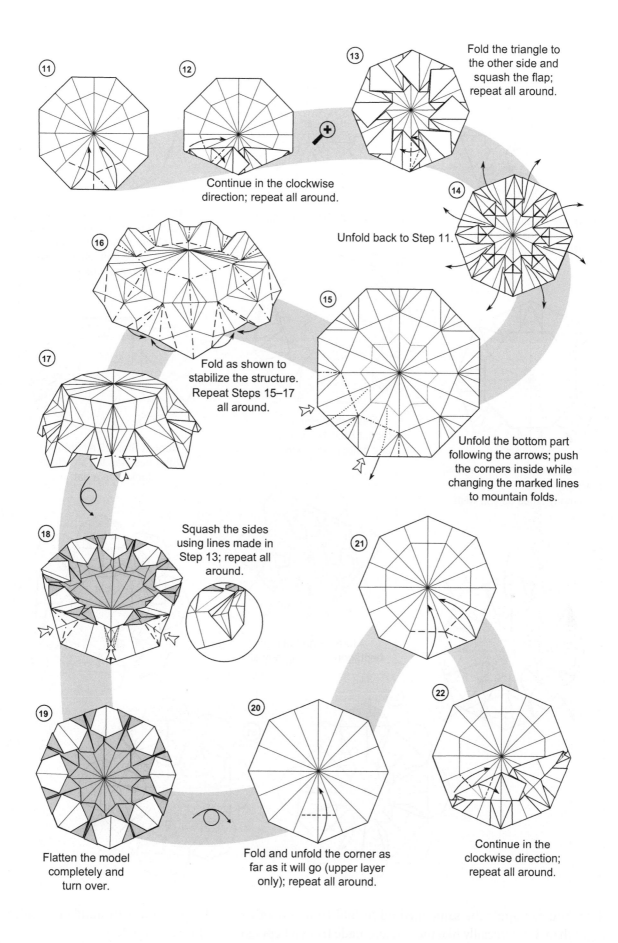

⑪

⑫

Continue in the clockwise
direction; repeat all around.

⑬ Fold the triangle to
the other side and
squash the flap;
repeat all around.

⑭ Unfold back to Step 11.

⑯ Fold as shown to
stabilize the structure.
Repeat Steps 15–17
all around.

⑮ Unfold the bottom part
following the arrows; push
the corners inside while
changing the marked lines
to mountain folds.

⑰

⑱ Squash the sides
using lines made in
Step 13; repeat all
around.

⑲ Flatten the model
completely and
turn over.

⑳ Fold and unfold the corner as
far as it will go (upper layer
only); repeat all around.

㉑

㉒ Continue in the
clockwise direction;
repeat all around.

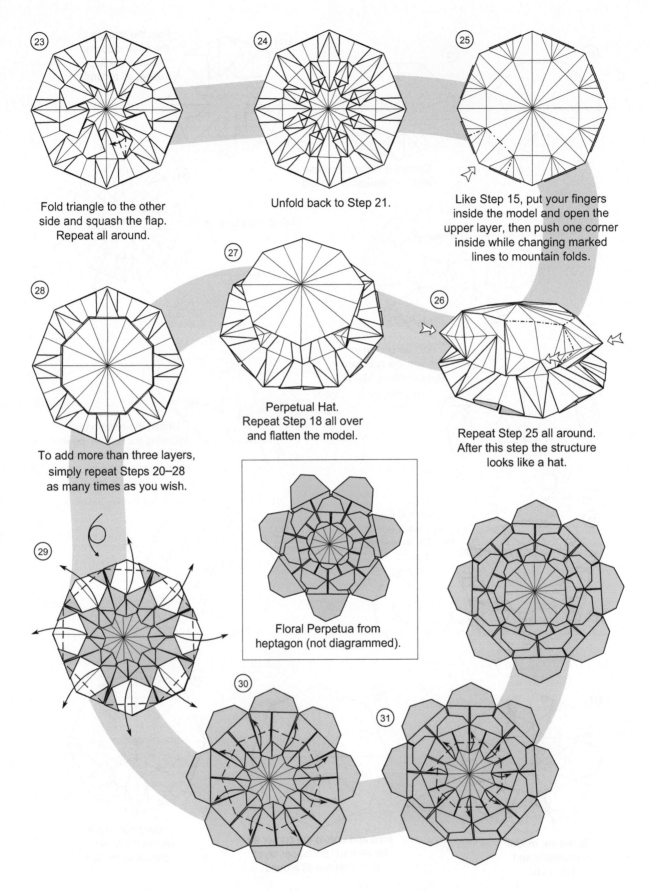

23 Fold triangle to the other side and squash the flap. Repeat all around.

24 Unfold back to Step 21.

25 Like Step 15, put your fingers inside the model and open the upper layer, then push one corner inside while changing marked lines to mountain folds.

27 Perpetual Hat. Repeat Step 18 all over and flatten the model.

26 Repeat Step 25 all around. After this step the structure looks like a hat.

28 To add more than three layers, simply repeat Steps 20–28 as many times as you wish.

Floral Perpetua from heptagon (not diagrammed).

29

30

31

Note: You can apply the same method to fold from any other regular polygon with number of sides larger than 5. I especially like the version made from a heptagon. Photos are on p. 120.

8 ◆ Decorative Boxes

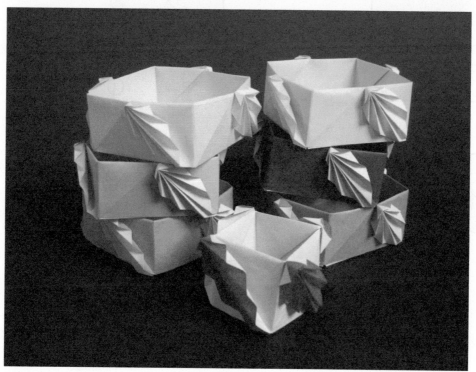

Top: Twisted Boxes and variations (next page). Bottom: Leafy Boxes (p. 135).

Twisted Box

(Created 2013 by Dáša Ševerová)

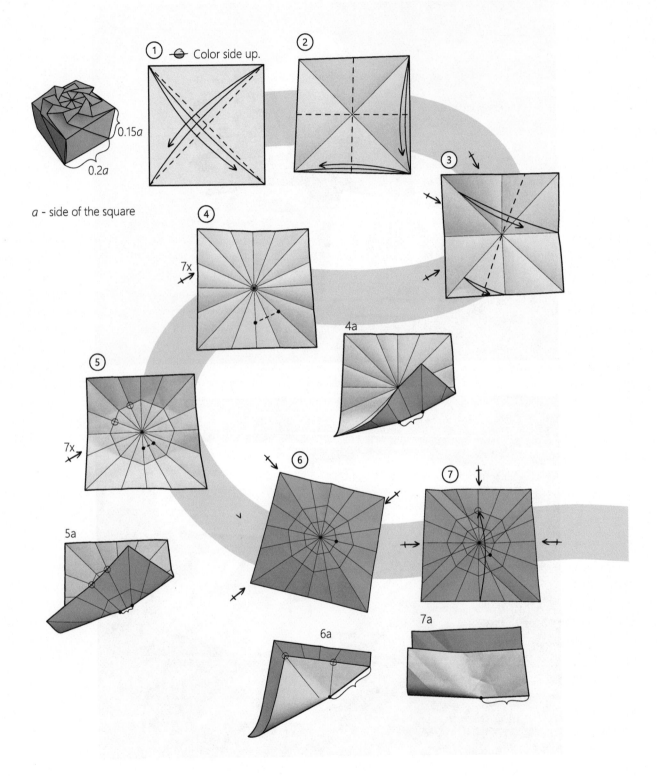

Color side up.

0.15a

0.2a

a - side of the square

7x

4a

7x

5a

6a

7a

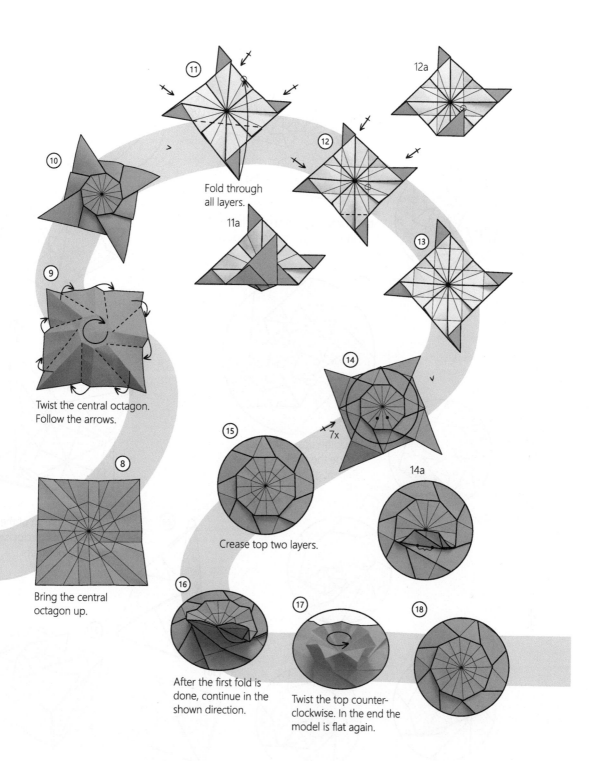

11 Fold through all layers.

11a

12a

12

13

10

9 Twist the central octagon. Follow the arrows.

14

15 Crease top two layers.

7x

14a

8 Bring the central octagon up.

16 After the first fold is done, continue in the shown direction.

17 Twist the top counter-clockwise. In the end the model is flat again.

18

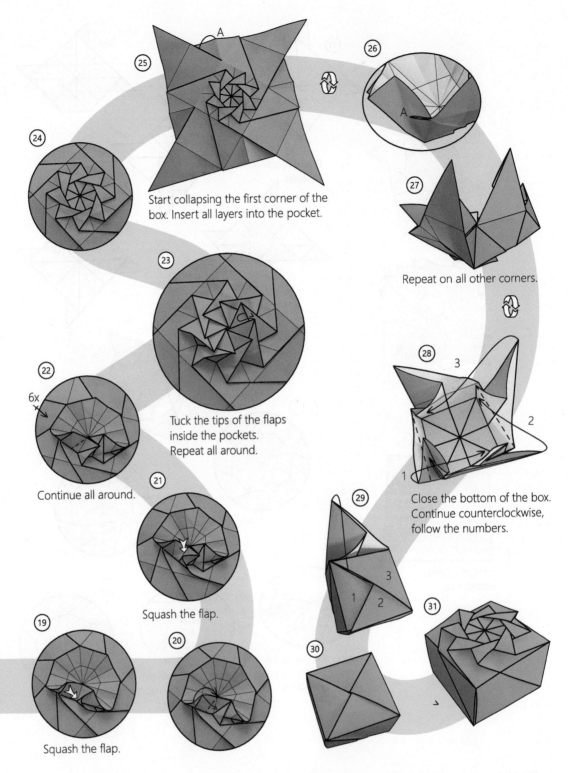

25 Start collapsing the first corner of the box. Insert all layers into the pocket.

26

27 Repeat on all other corners.

24

23 Tuck the tips of the flaps inside the pockets. Repeat all around.

22 6x Continue all around.

21 Squash the flap.

19 Squash the flap.

20

28 Close the bottom of the box. Continue counterclockwise, follow the numbers.

29

30

31

Note: Steps 14–24 could be replaced with other octagonal designs. Examples photo is on p. 131. These other possibilities are not diagrammed, but the reader is encouraged to experiment.

Leafy Box

Design and diagram © Christiane Bettens
www.origami-art.org

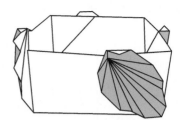

 (White side up)

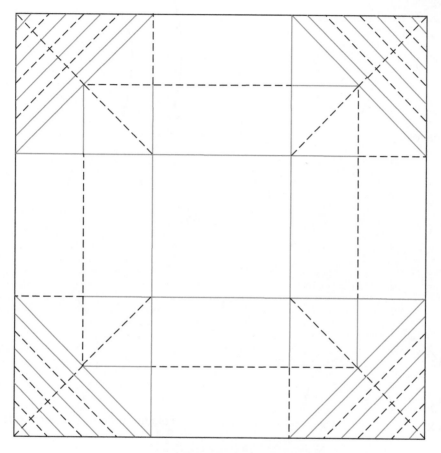

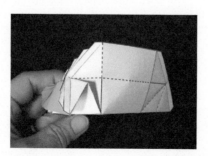

Note: For this design the author has provided a crease pattern and photo hint. If you have never folded from crease patterns, this is a great time to give it a try as this design is simple enough. Although it may appear challenging at first, you will succeed if you persevere.

Mountain folds are shown in red for clarity.

(Photo on p. 131.)

Four-Leaf Tato Box

Design and diagram © 2009 by Christiane Bettens, www.origami-art.org

1

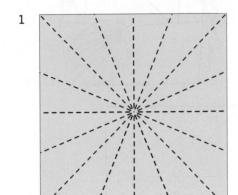

Color of leaves up

2

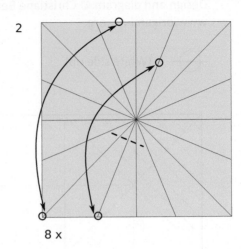

8 x

3

8 X   4 X

4a

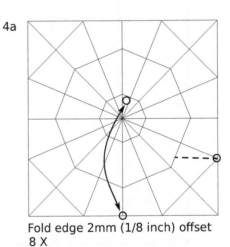

Fold edge 2mm (1/8 inch) offset
8 X

4b

Mountain fold from edge to outer octagon
8 X

5

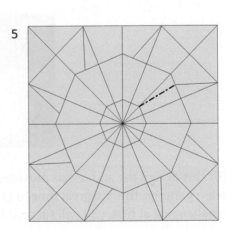

Extend mountain crease from step 4
to inner octagon
8 X

6

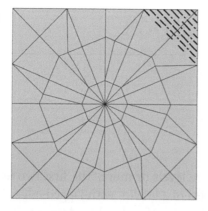

Accordion pleat corner. Repeat on other three corners.

7a

Start the collapse. This is the top view of the inside.

Side view

7b

Collapse in progress, side view.

Push flaps down to close box.

Two views of Four-Leaf Tato Boxes, top and side.

Hydrangea Box

(Created 2018)

Lid

Make Shuzo Fujimoto's Hydrangea [Fuj06, Fuj10], three levels or more. The first two levels will make the rim of the lid and the rest will be at the top. This idea of a lid rim with jagged edges is not new and has been independently folded by others. This example uses a five-level Hydrangea. You can also find video help in [Ada18].

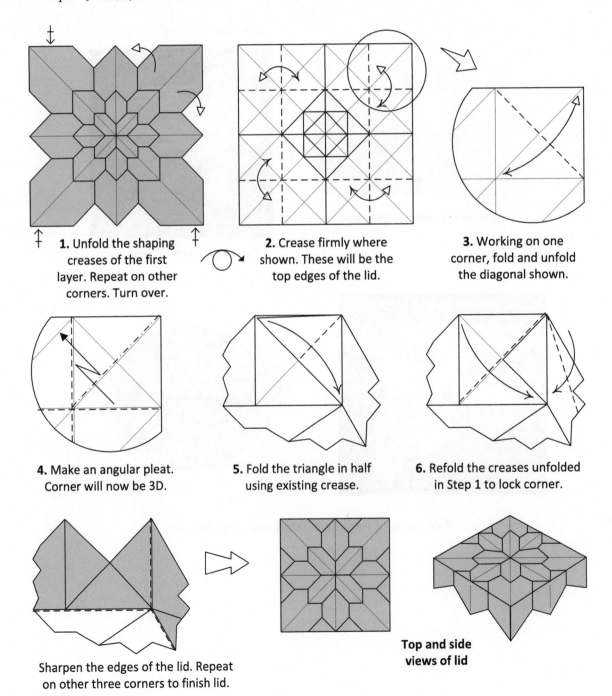

1. Unfold the shaping creases of the first layer. Repeat on other corners. Turn over.

2. Crease firmly where shown. These will be the top edges of the lid.

3. Working on one corner, fold and unfold the diagonal shown.

4. Make an angular pleat. Corner will now be 3D.

5. Fold the triangle in half using existing crease.

6. Refold the creases unfolded in Step 1 to lock corner.

Sharpen the edges of the lid. Repeat on other three corners to finish lid.

Top and side views of lid

Base

Use same size paper as lid. A contrasting color is recommended because it will show off the jagged edges of the lid better. We will fold like the traditional Masu Box [Kir05], but our proportions will be different so that we get a deeper base to show off the lid more prominently.

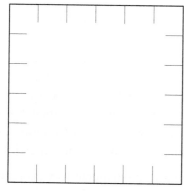

1. Pinch thirds both ways. Then pinch the sixths

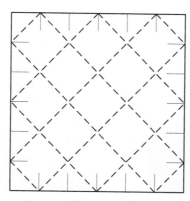

2. Crease the diagonals shown.

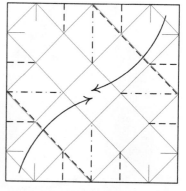

3. Extend some of the pinches up to the intersection of the diagonals. Refold two of the blintz folds.

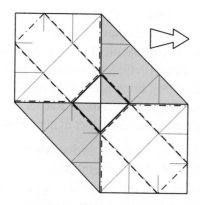

4. Collapse like a traditional Masu Box using existing creases. The box base will be cubic, i.e., of equal length, breadth, and height. The bolded square indicates bottom of the box.

Note: There are overlaps in the box base, which is not necessarily a drawback. It makes the box sturdy.

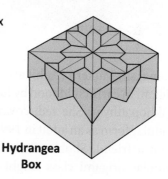

Hydrangea Box

Hydrangea Box with lid folded from a seven-level Hydrangea.

Author

Meenakshi Mukerji (Adhikari) was introduced to origami in early childhood by her uncle Bireshwar Mukhopadhyay. She rediscovered origami in its modular form as an adult in 1995, quite by chance, when a friend, Shobha Prabakar, took her to a modular origami class taught by Doug Philips. This newfound mathematical and structural side of modular origami rekindled her passion for the art, and soon after, she started designing and displaying origami on her popular website origamee. net. The website features colorful photo galleries and links to a myriad of free diagrams with nearly two million hits to date.

In 2005, Origami USA presented her with the Florence Temko award for generously sharing her work on her website. In April 2007, her first book *Marvelous Modular Origami* was published, followed in quick succession by six more books [Muk08, Muk10, Muk11, Muk13, Muk15, Muk18]. She has been a featured artist and a special guest at various origami conventions both in the USA and abroad. Although known for modular designs, she also has many single sheet designs to her credit.

Meenakshi regularly contributes to various origami periodicals and exhibits her work at conventions and various other exhibitions. She is a member of Origami USA and British Origami Society and an editor for OUSA's online magazine, *The Fold*. People who have provided her with much origami encouragement and inspiration include David Petty [Pet98], Rosalinda Sanchez, Robert Lang [Lan04], Francis Ow [Owrig], Rona Gurkewitz [Gurke], Ravi Apte, Rachel Katz [Kat01], and the numerous visitors of her website.

Born and raised in Kolkata, India, Meenakshi earned her BS in electrical engineering from the Indian Institute of Technology, Kharagpur, and her MS in computer science from Portland State University, Oregon. She then joined the software industry and worked for more than a decade. She is now at home in California devoting her time to family, traveling, designing origami, authoring origami books, and, of course, spreading the joy of origami.

Author's Website origamee.net

In 1997 the author created her first website on GeoCities (it later became Yahoo Geocities and is now no more), not only to showcase her work but equally to have an exercise in self-taught HTML. There was no social media then. The website then moved to another host for a few years but has been on the origamee.net domain since 2006. Since origami.net was not available, *origamee* was used to imply "Origami by Meenakshi." In

total, the website has been around for 20+ years with millions of visitors. The website is rich in photo galleries of the author's own designs and also of designs by others. You will find a plethora of free folding instructions as well. You can also follow the author on Flickr (https://www.flickr.com/photos/mmukhopadhyay/) and Facebook (https://www.facebook.com/origamee.net/).

Author's Previous Books

Books published by A K Peters/CRC Press

Marvelous Modular Origami (2007); Ornamental Origami:
Exploring 3D Geometric Designs (2008); Origami Inspirations (2010).

Self-Published Books

Exquisite Modular Origami (2011); Wondrous One Sheet Origami, First Edition
(2013, discontinued); Exquisite Modular Origami II (2015); Origami All Kinds (2018).

Contributors

Evan Zodl

with the world, he began making instructional origami videos on YouTube in 2008. Since that time, Evan (better known online as "EZ Origami" [Zod08]) has taught origami to millions of people around the globe through his videos.

On his channel (youtube.com/ezorigami), you will find detailed step-by-step tutorials for beginners and advanced folders. In addition to video tutorials, he has designed many unique origami models of his own. Much of his work is inspired by fractals, spirals, and recursive patterns found in nature. His original designs have been displayed in several notable exhibitions and published in various publications around the world. He currently works as a software engineer, continues to make video tutorials, and is in the process of publishing an origami book of his own.

Evan Zodl is a 23-year-old origami artist from New Jersey. He has been folding paper for over 14 years, and in an effort to share his passion

Dáša Ševerová

Dáša Ševerová is originally from Slovakia but currently lives in Switzerland. Dáša loves mathematics, especially geometry and solving puzzles, and

this is quite evident in her origami designs. Like many of us, she started doing origami as a child by folding some simple traditional models and paper airplanes and then forgot about it. Things changed while attending university when a friend showed her how to make different modular polyhedra from Penultimate Units [Plank] by Jim Plank. In her own words, "That's how my journey into the magical origami world really began." Soon after, she started creating her own beautiful designs. Not surprising, she prefers to create and fold geometric designs like flowers, stars, boxes, and polyhedra.

Dáša studied Mathematics and Biology for Teachers, and for several years she worked as a high school teacher in the same subject areas. She shares her origami designs and instructions on her Flickr gallery flickr.com/dasssa and recently more on her Instagram under nickname *paperain*. She attends origami conventions regularly and exhibits her designs. She was guest of honor at several conventions in Europe and recently also in USA. In May 2018 she published her first book, *Origami Journey* [Sev18] featuring geometric models of different kinds, which she created in last 10 years.

Christiane Bettens

Though Christiane Bettens studied and practices medicine in her hometown in Switzerland, her family has had a long relationship with paper involving editing, printing, and retailing. Ten years ago, her son asked if she could fold something other than a salt cellar. She folded the boat/hat, and while looking for more on the internet found the jumping frog and the flapping bird. These made her curiosity pique, following which she found pleasure in folding many of Tomoko Fuse's boxes and several variations of Tom Hull's PHiZZ modular units [Hul98, Hul12]. She focused on abstract, geometric models and was fascinated by being able to construct complicated structures with a simple piece of paper, without any measuring tools or calculations.

Christiane's design inspiration comes from creating variations of other's designs, identifying patterns in nature, experimenting with new techniques, and reverse-engineering designs from photos. She prefers the trial-and-error method without employing calculation or software. She has developed models in collaboration with friends and found the collective creating process fascinating. According to her, "I belong to a group of pioneers of tessellation. I'm glad the genre is now widespread with people finding their niche in it." She has also been fascinated with single sheet boxes and modular quilts. She goes by the nick name Mélisande* in her Flickr gallery flickr.com/melisande-origami and blog origami-art.org, which feature photos, crease patterns, and diagrams. She was special guest at Polish and French origami conventions and will be at an upcoming German convention. She regularly attends CDO (Italy), Origami USA (NY), and other conventions and exhibits internationally.

References

◈ [Ada07]Sara Adams, happyfolding.com, 2007

◈ [Ada14]Sara Adams, *Origami Instructions: Flower Gaillardia*, https://youtu.be/I8JOe1c3Dx8, 2014

◈ [Ada15]Sara Adams, *Mother's Day Origami Tutorial: Flower Hydrangea with Leaves*, https://youtu.be/TR-yYDvHxVE, 2015

◈ [Ada17]Sara Adams, *Origami Tutorial: Sunflower*, https://youtu.be/DoYy8Ux6qZw, 2017

◈ [Ada18]Sara Adams, *Mother's Day Origami Tutorial: Hydrangea Box*, https://youtu.be/HwDabCFU66o, 2018

◈ [Bet05]Christiane Bettens, *Mélisande*, flickr album, https://www.flickr.com/photos/melisande-origami/, 2005

◈ [Bet08]Christiane Bettens, *The Chronicles of Mélisande*, http://origami-art.org/blog/, 2008

◈ [Bos82]British Origami Society, *Convention Collection*, Autumn, 1982

◈ [Che14]Sy Chen, *Tridecagon*, http://freedomi.brinkster.net/Sy/Diagram/tridecagon.pdf, 2014

◈ [Dia06]Roman Diaz, flickr photostream, https://www.flickr.com/photos/88586913@N00, 2006

◈ [Dia12]Roman Diaz, Fractal Flower, *British Origami Magazine*, #275, pp. 26–34, 2012

◈ [Eng94]Peter Engel, *Folding the Universe: Origami from Angelfish to Zen*, Dover Publications, 1994

◈ [Eng11]Peter Engel, *Origami Odyssey: A Journey to the Edge of Paperfolding*, Tuttle Publishing, 2011

◈ [Eng16]Peter Engel, *10-Fold Origami*, Tuttle Publishing, 2016

◈ [Fuj06]Shuzo Fujimoto, *Hydrangea*, http://www.britishorigami.info/academic/johnsmith/hydrangea_john_smith.pdf, 2006

◈ [Fuj10]Shuzo Fujimoto, *Origami Project F—Hydrangea Folds*, Seibundo Shinkosha, 2010

◈ [Ger08]Robert Geretschläger, *Geometric Origami*, Arbelos, 2008

◈ [Gupta]Ishwat Gupta, Largest hexagon that can be inscribed within a square, https://www.geeksforgeeks.org/largest-hexagon-that-can-be-inscribed-within-a-square/

◈ [Gurke]Rona Gurkewitz, *Rona Gurkewitz' Modular Origami Polyhedra Systems Page*, http://make-origami.com/RonaGurkewitz/home.php

◈ [Hud11]Andrew Hudson, *Making a Successful Crease Pattern*, https://origamiusa.org/thefold/article/making-successful-crease-pattern-part-1, 2011

◈ [Hul98]Thomas Hull, *Origami Math*, http://mars.wne.edu/~thull/phzig/phzig.html, 1998–2000

◈ [Hul12]Thomas Hull, *Project Origami: Activities for Exploring Mathematics*, CRC Press, 2006

◈ [Jar05]Jorge Jaramillo, *Origami, flickr photostream*, https://www.flickr.com/photos/georigami/, 2005

◈ [Kat01]Rachel Katz, *Origami with Rachel Katz*, https://origamiwithrachelkatz.oriland.com/, 2001

◈ [Kaw70]Toyoaki Kawai, *Origami*, Hoikusha, 1970

◈ [Kom16]Hajime Komiya, *Ornamental Decoration Origami*, Lady Boutique Series, 2016

◈ [Kir05]Marc Kirschenbaum, *Masu Box Diagrams*, https://origamiusa.org/files/masu.pdf, 2005

◈ [Lan04]Robert J. Lang, http://langorigami.com, 2004

◈ [Lan15]Robert Lang, *Crease Patterns for Folders*, https://langorigami.com/article/crease-patterns-for-folders/, 2015

◈ [Lan18]Robert Lang, *Twists, Tilings, and Tessellations*, CRC Press, 2018

◈ [Mat01]Mathematics Stack Exchange, Regular pentagon in a square, https://math.stackexchange.com/questions/1886807/regular-pentagon-in-a-square/, 2016

◈ [Mon12]John Montroll, *Galaxy of Origami Stars*, self-published, 2012

◈ [Muk97]Meenakshi Mukerji, *Origami by Meenakshi*, http://www.origamee.net, 1997

◈ [Muk07]Meenakshi Mukerji, *Marvelous Modular Origami*, CRC Press, 2007

◈ [Muk08]Meenakshi Mukerji, *Ornamental Origami: Exploring 3D Geometric Designs*, CRC Press, 2008

◈ [Muk10]Meenakshi Mukerji, *Origami Inspirations*, CRC Press, 2010

◈ [Muk11]Meenakshi Mukerji, *Exquisite Modular Origami*, self-published, 2011

◈ [Muk13]Meenakshi Mukerji, *Wondrous One Sheet Origami*, First Edition, self-published, 2013

◈ [Muk15]Meenakshi Mukerji, *Exquisite Modular Origami II*, self-published, 2015

◈ [Muk18]Meenakshi Mukerji, *Origami All Kinds*, self-published, 2018

◈ [Noa17]NOA, *Origami Symposium Diagram Collection*, 2017

◈ [ORC19]Origami Resource Center, *Crease Patterns*, https://www.origami-resource-center.com/crease-patterns.html, 2019

◈ [Owrig]Francis Ow, *Owrigami*, http://owrigami.com/

◈ [Pam08]Chris K. Palmer, *Shadowfolds*, http://shadowfolds.com/, 2008

◈ [Pet98]David Petty, http://www.davidpetty.me.uk/, 1998

◈ [Plank]James S. Plank, *Jim Plank's Origami Page (Modular)* http://web.eecs.utk.edu/~jplank/plank/origami/

◈ [Sev06]Dáša Ševerová, *Origami, flickr photostream*, https://www.flickr.com/photos/dasssa, 2006

◈ [Sev18]Dáša Ševerová, *Origami Journey: Into the Fascinating World of Geometric Origami*, self-published, 2018

◈ [Sha02]Jeremy Shafer, *Origami to Befriend and Befuddle*, BARF Newsletter, 2002

◈ [Som07]Endre Somos, flickr photostream, https://www.flickr.com/photos/44259004@N00/, 2007

◈ [Yos84]Akira Yoshizawa, *Sosaku Origami—Creative Origami*, NHK, 1984

◈ [Zod08]Evan Zodl, *EZ Origami*, https://ez-origami.com/, 2008

Index

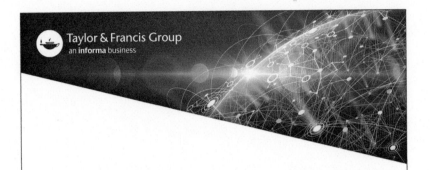